MAJOR
APPLIANCES

Other Publications:

FITNESS, HEALTH & NUTRITION
SUCCESSFUL PARENTING
HEALTHY HOME COOKING
UNDERSTANDING COMPUTERS
THE ENCHANTED WORLD
THE KODAK LIBRARY OF CREATIVE PHOTOGRAPHY
GREAT MEALS IN MINUTES
THE CIVIL WAR
PLANET EARTH
COLLECTOR'S LIBRARY OF THE CIVIL WAR
THE EPIC OF FLIGHT
THE GOOD COOK
WORLD WAR II
HOME REPAIR AND IMPROVEMENT
THE OLD WEST

MAJOR APPLIANCES

TIME-LIFE BOOKS
ALEXANDRIA, VIRGINIA

Fix It Yourself was produced by
ST. REMY PRESS

MANAGING EDITOR	Kenneth Winchester
MANAGING ART DIRECTOR	Pierre Léveillé

Staff for *Major Appliances*

Editors	Kathleen M. Kiely, Randall Duckett
Art Director	Diane Denoncourt
Research Editors	Nancy D. Kingsbury, Elizabeth W. Lewis
Contributing Writers	Cathleen Farrell, Michèle McLaughlin, Jeremy Searle, Dianne Thomas
Contributing Illustrators	Gérard Mariscalchi, Jacques Proulx
Technical Illustrator	Robert Paquet
Cover	Robert Monté
Index	Christine M. Jacobs
Administrator	Denise Rainville
Coordinator	Michelle Turbide
Systems Manager	Shirley Grynspan
Systems Analyst	Simon Lapierre
Studio Director	Daniel Bazinet

Time-Life Books Inc. is a wholly owned subsidiary of
TIME INCORPORATED

FOUNDER	Henry R. Luce 1898-1967
Editor-in-Chief	Henry Anatole Grunwald
Chairman and Chief Executive Officer	J. Richard Munro
President and Chief Operating Officer	N. J. Nicholas Jr.
Chairman of the Executive Commitee	Ralph P. Davidson
Corporate Editor	Ray Cave
Group Vice President, Books	Reginald K. Brack Jr.
Vice President, Books	George Artandi

TIME-LIFE BOOKS INC.

EDITOR	George Constable
Director of Design	Louis Klein
Director of Editorial Resources	Phyllis K. Wise
Acting Text Director	Ellen Phillips
Editorial Board	Russell B. Adams Jr., Dale M. Brown, Roberta Conlan, Thomas H. Flaherty, Donia Ann Steele, Rosalind Stubenberg, Kit van Tulleken, Henry Woodhead
Director of Photography and Research	John Conrad Weiser
PRESIDENT	Christopher T. Linen
Executive Vice President	John M. Fahey Jr.
Senior Vice President	James L. Mercer, Leopoldo Toralballa
Vice Presidents	Stephen L. Bair, Ralph J. Cuomo, Terence J. Furlong, Neal Goff, Stephen L. Goldstein, Juanita T. James, Hallett Johnson III, Robert H. Smith, Paul R. Stewart
Director of Production Services	Robert J. Passantino

Editorial Operations

Copy Chief	Diane Ullius
Editorial Operations	Caroline A. Boubin (manager)
Production	Celia Beattie
Quality Control	James J. Cox (director)
Library	Louise D. Forstall
Correspondents	Elisabeth Kraemer-Singh (Bonn); Maria Vincenza Aloisi (Paris); Ann Natanson (Rome).

THE CONSULTANTS

Consulting Editor **David L. Harrison** is Managing Editor of Bibliographics Inc. in Alexandria, Virginia. He served as an editor of several Time-Life Books do-it-yourself series, including *Home Repair and Improvement, The Encyclopedia of Gardening* and *The Art of Sewing.*

John and **Jeff Lefever** are operators of Alco Appliance Inc. in Beltsville, Maryland. The Lefever brothers have worked for many years repairing appliances and managing appliance-repair companies.

Evan Powell is the Director of Chestnut Mountain Research Inc. in Taylors, South Carolina, a firm specializing in the development and evaluation of home appliances. He is a contributing editor for several do-it-yourself magazines, and the author of two books on home repair.

Oleh Z. Kowalchuk and **Richard Thibault**, special consultants for Canada, are National Product Specialists with CAMCO, a major appliance manufacturer. Trained as technicians, they have written service manuals and now train field service personnel.

Bruce Barcomb repairs major appliances for Barcomb's TV and Furniture Center in Champlain, New York.

Library of Congress Cataloguing in Publication Data
Major appliances
(Fix it yourself)
Includes index.
1. Household appliances, Electric—Maintenance and repair—Amateurs's manuals. I. Time-Life Books. II. Series.
TK9901. M28 1987   683'.83   86-30136
ISBN 0-8094-6204-4
ISBN 0-8094-6205-2   (lib. bdg.)

For information about any Time-Life book, please write:
Reader Information
541 North Fairbanks Court
Chicago, Illinois 60611

CONTENTS

HOW TO USE THIS BOOK

Major Appliances is divided into three sections. The Emergency Guide on pages 8-13 provides information that can be indispensable, even lifesaving, in the event of a household emergency. Take the time to study this section *before* you need the important advice it contains.

The Repairs section—the heart of the book—is more than a collection of how-to tips and techniques. It is a system for troubleshooting and repairing refrigerators, freezers, ranges, dishwashers, garbage disposers, clothes washers and dryers. Pictured below are four sample pages from the Dishwashers chapter, with captions describing the various features of the book and how they work. If your dishes are dirty or spotted after washing, for example, the Troubleshooting Guide will offer a number of possible causes ranging from incorrect loading of dishes to a faulty timer. If the problem is a clogged spray arm, you will be directed to page 80 for detailed, step-by-step directions for servicing the spray arm and tower.

Each job has been rated by degree of difficulty and the average time it will take for a do-it-yourselfer to complete. Keep in mind that this rating is only a suggestion. Before deciding whether you should attempt a repair, first read all the instructions carefully. Then be guided by your own confidence, and

---

Introductory text
Describes proper use and care of the appliance, most common breakdowns and basic safety precautions.

"Exploded" and cutaway diagrams
Locate and describe the various mechanical, electrical and plumbing parts of the appliance.

Troubleshooting Guide
To use this chart, locate the symptom that most closely resembles your appliance problem, review the possible causes in column 2, then follow the recommended procedures in column 3. Simple fixes may be explained on the chart; in most cases you will be directed to an illustrated, step-by-step repair sequence.

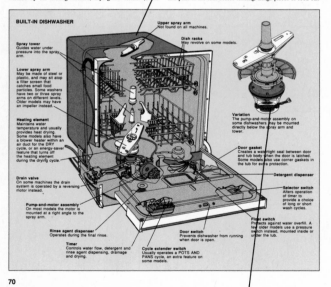

Variations
Differences in popular appliance models are described throughout the book, particularly if a repair procedure varies from one machine to another, or from electric to gas models.

Degree of difficulty and time
Rate the complexity of each repair, and how much time the job should take for a homeowner with average do-it-yourself skills.

Special tool required
Some electrical repairs, particularly those involving heating elements, pumps or motors, require a multitester *(page 131)*.

the tools and time available to you. For complex or time-consuming repairs, such as replacing a motor, you may wish to call for professional service. You will still have saved time and money by diagnosing the problem yourself.

Most of the repairs in *Major Appliances* can be made with screwdrivers, wrenches and utility pliers. Some troubleshooting procedures require either a continuity tester or the more precise multitester. Basic appliance repair tools—and the proper way to use them—are presented in the Tools & Techniques section starting on page 130. If you are a novice when it comes to home repair, read this section in preparation for a major job.

Home repair can lead to serious injury unless you take certain basic precautions. Before working on an appliance, always unplug its power cord or shut off power at the main service panel *(page 132)*. If you must plug in an appliance to test it, always unplug it again before proceeding. Do not disconnect or move a gas appliance yourself; call the gas company or a service technician to do it for you. When removing large panels, wear gloves or tape sharp metal edges with masking tape. Keep a bucket and rags handy when working with hoses, valves and pumps. Most important, follow all safety tips and **Caution** warnings throughout the book.

Name of repair
You will be referred by the Troubleshooting Guide to the first page of a specific repair job.

Step-by-step procedures
Follow the numbered repair sequence carefully. Depending on the result of each step, you may be directed to a later step, or to another part of the book, to complete the repair.

Tools and techniques
When a tool or method is required for a job, it is described within the step-by-step repair. General information on working with appliances, including the use of continuity testers and multitesters, is covered in the Tools & Techniques section *(page 130)*.

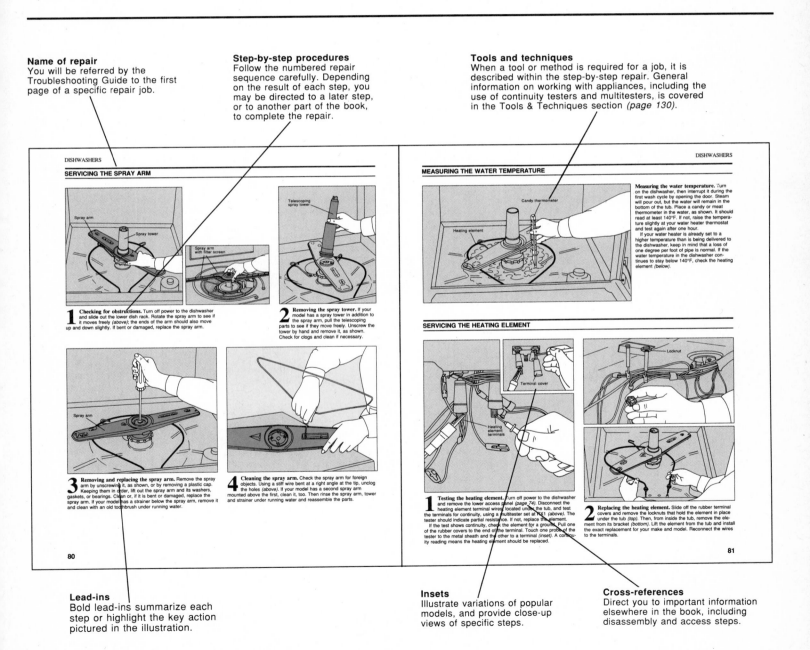

Lead-ins
Bold lead-ins summarize each step or highlight the key action pictured in the illustration.

Insets
Illustrate variations of popular models, and provide close-up views of specific steps.

Cross-references
Direct you to important information elsewhere in the book, including disassembly and access steps.

EMERGENCY GUIDE

Preventing appliance problems. A faulty appliance usually creates a minor, temporary annoyance—the icemaker won't make ice, or the clothes washer doesn't drain. On rare occasions, however, mechanical failure in an appliance, coupled with your home's electrical, gas or plumbing systems, can cause a real hazard.

You can prevent most appliance-related mishaps by exercising the commonsense precautions presented here. The key to appliance safety is proper use and regular maintenance of the machines themselves, and the household utilities that power them. Eager to help you use their products safely, appliance manufacturers provide a Use and Care manual with every new appliance. Several major manufacturers offer toll-free consumer "hot lines" for answering questions immediately, and they will mail you a Use and Care manual free upon request. Their telephone numbers are available from the 800 operator.

The repair of major appliances need not be any more dangerous than their daily use. In fact, proper repairs will prevent hazardous conditions caused by wear and neglect. The list of safety tips at right covers basic guidelines for safe service and use of any appliance; see the chapters on individual appliances for more specific advice.

Accidents may befall the most careful of households. Fire, the most insidious, can be deprived of its sneak attack by judiciously placed smoke alarms, and a fire extinguisher can snuff the blaze before it gets the upper hand. Learn how to choose and use these safety tools on pages 10 and 11.

The Troubleshooting Guide on page 9 puts emergency procedures at your fingertips. It lists the quick-action steps to take, and refers you to the procedures on pages 10 through 13 for more detailed instructions. Read the emergency instructions thoroughly before you need to use them. And familiarize yourself with the Tools and Techniques section *(page 130)*, which covers your home's electrical, gas and plumbing systems.

When in doubt about the safety of an appliance or your ability to handle an emergency, don't hesitate to call for help. Post the telephone numbers for the fire department, gas company and electric company near the telephone. Even in non-emergency situations, they can answer questions concerning the proper use of appliances and household utilities.

SAFETY TIPS

1. Familiarize yourself with the Use and Care manual for each appliance. If you have misplaced the manual, purchase one from a parts distributor, or call the manufacturer's toll-free number for a new copy.

2. Before attempting any repair in this book, read the entire repair procedure. Familiarize yourself with the specific safety information presented for each appliance.

3. Let a heating appliance, such as a dryer or range, cool completely before starting repairs.

4. Before servicing an appliance, unplug the power cord or disconnect power at the service panel *(p. 132)*. Leave a note on the panel so that no one reconnects the power while you are working.

5. Label your electric service panel with the locations of the appliance circuit breakers or fuses.

6. Light the work area well, and do not reach into any area of the machine you cannot see clearly.

7. Beware of sharp metal edges and pointed screws; pad them with masking tape to avoid cutting yourself.

8. Never bypass or alter any appliance switch or component. Do not remove the ground plug of a 3-prong power cord.

9. Use only replacement parts and wiring of the same specifications as the original part. If in doubt, ask the appliance parts dealer.

10. When reassembling an appliance, take care that all wire connections are secure, and that no wires are pinched between panels or moving parts.

11. Completely reassemble an appliance before reconnecting the power, water or gas supply.

12. If in doubt about the safety of an appliance, call a professional service technician.

13. Label the shutoff valves for gas appliances, as well as the house's main gas valve.

14. Never light a flame while working on a gas appliance.

15. Know where the water shutoff valves are for the washer, dishwasher and icemaker, as well as the house's main water shutoff valve. Label them.

16. If hot water is not used for two weeks or more, hydrogen can build up in the hot-water heater and pipes. Before turning on appliances that use water, run all hot water taps in the house for two minutes to clear out the gas.

17. Post emergency, utility company and repair service numbers near the telephone.

18. Install smoke detectors and fire extinguishers in your home *(p. 11)*.

19. Do not allow children to play on or operate appliances.

TROUBLESHOOTING GUIDE

PROBLEM	PROCEDURE
FIRE EMERGENCIES	
Fire in an electrical outlet	Use fire extinguisher *(p. 10)*
	Disconnect power at service panel, then unplug power cord *(p. 11)*
	If flames or smoldering continue, leave house and call fire department
Cooking fire in a pan on range top	Do not move pan
	Turn off burners and exhaust hood fan
	Slide fitted lid onto pan *(p. 10)*
Cooking fire spread to range top	Turn off burners and exhaust hood fan
	Pour baking soda or salt on fire to smother it *(p. 10)*
	Do not apply water, baking powder, flour or other household substances to fire
	If flames spread, use fire extinguisher *(p. 10)*
Cooking fire in oven	Close oven door to smother flames; turn off oven and allow it to cool
	If flames spread, use fire extinguisher *(p. 10)*
Grease fire in range hood	Turn off range hood fan
	If flames do not subside within a few seconds, use fire extinguisher *(p. 10)*
Burning smell or smoke from appliance	Do not open appliance door
	Unplug power cord *(p. 11)* and allow appliance to cool
	If flames develop, use fire extinguisher *(p. 10)*
Appliance on fire	Use fire extinguisher *(p. 10)*
	Disconnect power at service panel *(p. 11)*, then unplug power cord *(p. 11)*
Clothes burning or melting in dryer	Close dryer door to smother fire
	Shut off power *(p. 11)*; allow dryer to cool before removing clothes
	If flames develop, use fire extinguisher *(p. 10)*
ELECTRICAL EMERGENCIES	
Appliance gives off sparks or shocks user	Unplug power cord *(p. 11)* without touching appliance, or disconnect power at service panel *(p. 11)*
Power cord plug and wall outlet sparking, discolored, hot to the touch or melting	Disconnect power at service panel by tripping circuit breaker or removing fuse *(p. 11)*
	Unplug power cord *(p. 11)*
	Call an electrician to inspect outlet and wiring for damage
GAS EMERGENCIES	
Pilot light out in gas appliance	Relight pilot *(p. 12)*
Odor of escaping gas	Ventilate room *(p. 12)*
	Do not touch electrical outlets or switches
	Extinguish all flames
	Check pilots of all gas appliances and relight if necessary *(p. 12)*
Persistent odor of gas with all pilots lit	Ventilate room *(p. 12)*
	Turn off gas supply valve to all gas appliances *(p. 12)*
	Leave house and call gas company
WATER EMERGENCIES	
Clothes washer overflowing	Turn off washer
	Set timer on final spin cycle and turn on washer to pump out water
	If water doesn't drain: Unplug power cord *(p. 12)* and bail or siphon out water *(p. 13)*
Dishwasher overflowing	Turn off dishwasher at timer
	Turn off dishwasher water valve under sink *(p. 13)*
	Turn on dishwasher and let run to empty water
	If water doesn't drain: Shut off power *(p. 11)* and bail or siphon out water *(p. 13)*
Appliance leaking	Turn off machine
	Turn off water supply to machine *(p. 13)*
	If machine is full of water: Shut off power *(p. 11)* and bail or siphon out water *(p. 13)*
Icemaker doesn't stop filling with water	Turn off icemaker water supply valve, usually under sink *(p. 13)*
Water on floor from appliance leak or overflow	If you must stand in water to mop it up, first unplug power cord or disconnect power *(p. 11)*
	Dam area around water with washable, absorbent rugs; clean up with mop or towels *(p. 13)*
Appliance or electrical outlet submerged	Do not enter room; disconnect power at service panel *(p. 11)*

FIRE EMERGENCIES

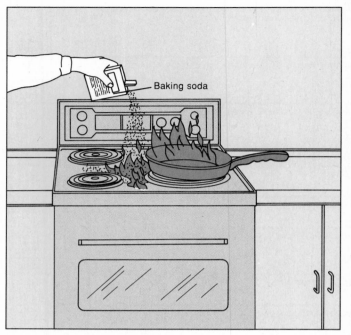

Smothering a cooking fire in a pan. Do not move the pan. Turn off the burner controls and the range hood fan. Protecting your hand with an oven mitt or a pot holder, slide a fitted lid onto the pan, as shown. If no lid is available, use a plate or platter a bit larger than the pan. Do not clap the cover straight down; the rush of air can spread the flames. Let the pan cool before removing the cover. If the fire spreads, apply baking soda *(next step)*.

Controlling a cooking fire on the range top. Do not move the pan. Turn off the burner controls and the range hood fan. Pour baking soda liberally over the flames until they are out *(above)*. If baking soda is not available, use salt. **Caution:** Do not put water, baking powder or flour on a cooking fire; they will spread the flames. Allow the range to cool before removing the pan and cleaning the range top. If the fire spreads, use a fire extinguisher *(next step)*.

Using a fire extinguisher. To snuff an oven fire, first turn off the oven control and close the door; lack of oxygen should kill the flames. If a range fire—or any appliance fire—spreads, use a fire extinguisher rated ABC or BC. Stand near an exit, 6 to 10 feet from the fire. Pull the lock pin out of the extinguisher handle and, holding the extinguisher upright, aim the nozzle at the base of the flames. Squeeze the two levers of the handle together, spraying in a quick side-to-side motion. The fire may flare and appear to grow at first before subsiding. If the discharge stream scatters the flames, move back. Keep spraying until the fire is completely extinguished. Watch carefully for "flashback," or rekindling, and be prepared to spray again. Allow the appliance to cool completely before cleaning it.

SAFETY ACCESSORIES

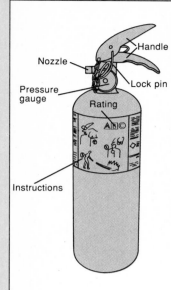
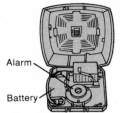
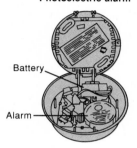

A household fire extinguisher. Best for use in the home is a multipurpose dry-chemical extinguisher rated ABC *(left)*, which is safe and effective against fires in upholstery, kitchens and electrical appliances alike. An extinguisher of convenient size holds a pressurized cargo of 2 1/2 to 7 pounds. Check the pressure gauge monthly; after any discharge or loss of pressure, have the tank recharged professionally or buy a new extinguisher. Mount extinguishers, using the wall bracket provided, near doors to the kitchen, utility room, garage and basement.

Nozzle
Handle
Pressure gauge
Lock pin
Rating
ABC
Instructions

Ionization alarm
Alarm
Battery

Photoelectric alarm
Battery
Alarm

Two kinds of smoke alarms. Ionization alarms *(left, top)*, which sense atomic particles, respond quickly to hot fires with little smoke, but they tend to set off annoying false alarms in the presence of normal cooking fumes.

Photoelectric alarms *(left, bottom)* "see" smoke molecules; they respond best to the smoldering typical of cooking, appliance and upholstery fires.

Install at least one smoke alarm in a central hallway on every floor of the house, near the kitchen, bedrooms and head of the stairs, as well as in the garage and basement. Mount a battery-powered smoke alarm on the ceiling. Replace the battery once a year—the detector emits a chirping sound when the battery runs low.

ELECTRICAL EMERGENCIES

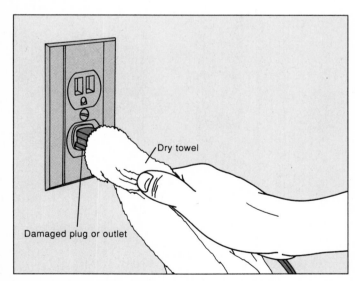

Dry towel
Damaged plug or outlet

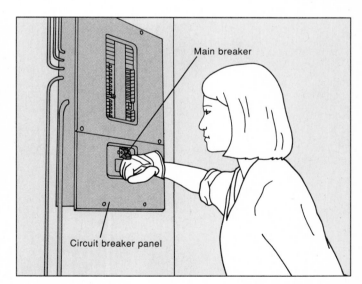

Main breaker
Circuit breaker panel

Pulling the power cord. Caution: If you are standing on a wet floor, or the outlet or plug is sparking, burning or melting, do not touch the power cord; instead, disconnect power at the service panel *(next step)*. If the appliance sparks, shocks you, feels abnormally hot or is burning, disconnect the power cord. Protect your hand with a thick, dry towel or a heavy work glove. Without touching the machine, grasp the cord several inches from the plug and pull it out *(above)*.

Disconnecting power at the service panel. If the floor is wet, or the appliance outlet is burning or sparking, wear heavy, dry gloves and put one hand behind your back. Flip off the breaker, or unscrew the fuse, that controls the appliance's circuit. (A dryer or range may have two breakers or fuses.) If you are unsure of the circuit, shut off the main power breaker. Use the back of your hand *(above)*; any shock will jerk your hand away from the panel. To disconnect a main fuse block, insert a dry wooden stick behind the handle and pull it out.

GAS EMERGENCIES

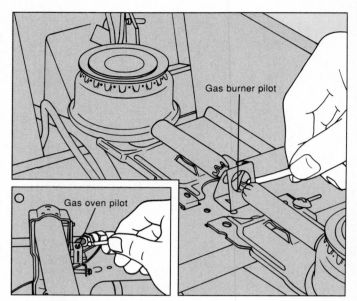

Gas burner pilot

Gas oven pilot

1 **Ventilating a gas-filled room.** First open all windows and doors in the room *(above)*. **Caution:** Extinguish all flames, except for pilot lights. Do not use electrical switches or outlets—a spark could ignite the gas. Turn off controls to all gas appliances in the room. When the gas has dissipated, relight any pilots that are out *(step 2)*. If the gas does not dissipate, leave the house and call the gas company.

2 **Relighting a pilot.** Check all the pilots in every gas appliance in the house to find the pilots that are out. (If you don't know where the pilots are, consult the appliance's Use and Care manual or the access instructions in this book.) Wait a minute for gas in the appliance to dissipate, then light the pilot with a match as shown for a gas range *(above)* or oven *(inset)*. If the pilot does not light or stay lit, clean or adjust it *(page 60)* and try again. If you smell gas with all pilots lit, or in an appliance with electronic ignition, turn off the gas supply *(step 3)*.

Gas supply pipe

Shutoff valve

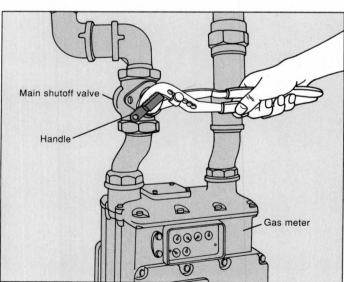

Main shutoff valve

Handle

Gas meter

3 **Turning off the gas supply.** If the appliance has a valve on its gas supply pipe, turn the handle perpendicular to the pipe to shut off the gas *(above, left)*. If the gas in the room does not dissipate, leave the house and call the gas company. The gas supply to the whole house may be turned off at the meter; using a wrench, turn the main valve so that its handle is perpendicular to the pipe *(above, right)*. Leave the house and call the gas company.

WATER EMERGENCIES

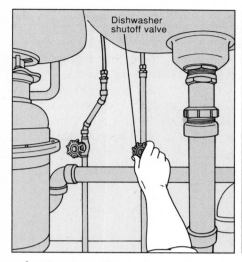

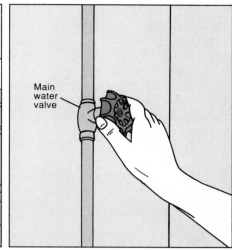

1 **Turning off the water supply. Caution:** If an electrical appliance is submerged, do not enter the room; disconnect power at the service panel *(page 11)*. If an appliance is leaking or overflowing, unplug the power cord *(page 11)*, then turn off the water at the valve to the appliance. For a dishwasher, the valve is usually under the sink *(above, left)*. Clothes washer valves are on the wall behind the machine *(above, center)* or at the utility sink faucet; shut off both the hot and cold water supply. If the valve is leaking, or there is no valve, turn off the house water supply at the main valve *(above, right)*, located near the water meter or where the main water supply pipe enters the house.

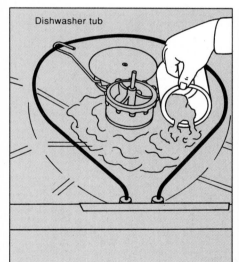

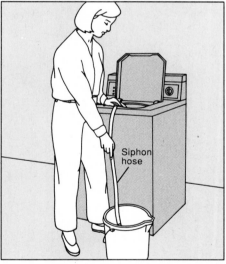

2 **Emptying an appliance of water.** If a clothes washer stops working in the wash or rinse cycle, try setting it on the spin cycle to drain the water. To empty an overflowing dishwasher, turn off the water valve under the sink *(step 1)* and turn on the dishwasher. If a water-filled appliance does not work at all, unplug it and let the water cool. Bail out the water *(above, left)*, or use a hose to siphon it out into a bucket on the floor *(above, right)*. To start the water flow, suck on the end of the hose; the flow will continue as long as the end of the hose is lower than the level of the water.

3 **Damming a leak.** Disconnect power to the appliance without stepping in the water *(page 11)*. To keep water on the floor from spreading, surround it with a dam of washable, rolled-up rugs or towels. Clean up water within the dammed area with a mop or heavy cotton towels, not paper towels.

REFRIGERATORS

An average lifespan of fifteen years puts refrigerators among the longest lasting and most trouble-free of all major appliances. All refrigerators work by means of a sealed cooling system. Refrigerant gas, liquefied by high pressure, passes through a narrow capillary tube and enters the evaporator coils inside the refrigerator. Under reduced pressure, it quickly boils into a gas, absorbing heat in the process. This gas flows to the compressor, which pumps it into the condenser coils on the outside of the refrigerator. Now under high pressure, the refrigerant gives off heat to the surrounding air as it returns to a liquid state. The refrigerant then passes back through the capillary tube into the evaporator coils as the cycle of heating and cooling continues. When the desired temperature is reached, the thermostat control turns off the compressor.

The sealed cooling system limits the range of jobs that can be tackled by a do-it-yourselfer. Repairs to the compressor, evaporator or condenser require special skills and tools and must be handled by professionals. But many other problems can be diagnosed and repaired without any special tools.

Though they chill food the same way, refrigerators vary in style and features. Shown below is a typical two-door, frost-free model with floor-level condenser coils. Variations are described within each repair step. Follow the instructions that most closely fit your machine.

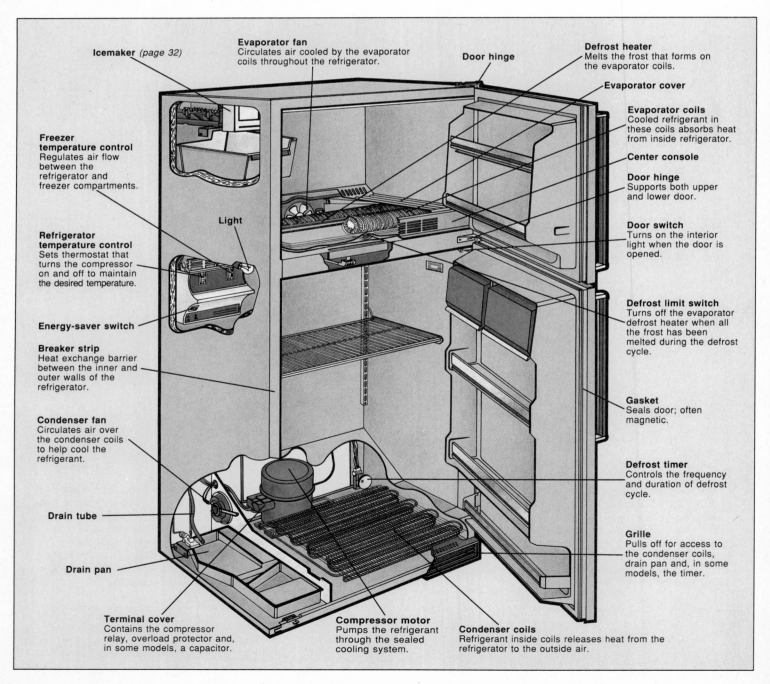

Icemaker *(page 32)*

Evaporator fan
Circulates air cooled by the evaporator coils throughout the refrigerator.

Door hinge

Defrost heater
Melts the frost that forms on the evaporator coils.

Evaporator cover

Evaporator coils
Cooled refrigerant in these coils absorbs heat from inside refrigerator.

Center console

Door hinge
Supports both upper and lower door.

Door switch
Turns on the interior light when the door is opened.

Freezer temperature control
Regulates air flow between the refrigerator and freezer compartments.

Light

Refrigerator temperature control
Sets thermostat that turns the compressor on and off to maintain the desired temperature.

Energy-saver switch

Breaker strip
Heat exchange barrier between the inner and outer walls of the refrigerator.

Condenser fan
Circulates air over the condenser coils to help cool the refrigerant.

Defrost limit switch
Turns off the evaporator defrost heater when all the frost has been melted during the defrost cycle.

Gasket
Seals door; often magnetic.

Defrost timer
Controls the frequency and duration of defrost cycle.

Drain tube

Drain pan

Grille
Pulls off for access to the condenser coils, drain pan and, in some models, the timer.

Terminal cover
Contains the compressor relay, overload protector and, in some models, a capacitor.

Compressor motor
Pumps the refrigerant through the sealed cooling system.

Condenser coils
Refrigerant inside coils releases heat from the refrigerator to the outside air.

Refrigerators vary mainly in the way they handle the home-maker's bugaboo: frost buildup. A frost-free refrigerator has a defrost heater that warms the evaporator coils in both the refrigerator and freezer compartments. Pictured throughout this chapter, it contains all of the components found in the other two types. A semi-automatic refrigerator has a defrost heater that prevents frost in the refrigerator compartment only; the freezer compartment must be defrosted by hand. A manual refrigerator has no defrost heater and must also be defrosted by hand. To defrost a refrigerator, turn off or unplug the machine and use a hair dryer on a low setting to melt a light layer of frost. For a heavy buildup, place a pan of hot water in each compartment and close the doors. Never hammer at ice buildup or try to pry it away with a sharp tool; this may puncture the evaporator.

Regular cleaning prolongs the life of a refrigerator and keeps energy costs down. Brush or vacuum the condenser coils at the bottom or back of the machine regularly *(page 17)*. If the coils are greasy, wash them with soapy water. To prevent odors and ice, keep the drain tubes and drain pan clean and unblocked. Using an oven baster, flush the drain tubes with warm water and baking soda *(page 18)*. Wash the compartments, trays, shelves and drain pan twice a year using a water and baking soda solution followed by a clear water rinse.

If the power goes out, food will keep in a closed refrigerator for 24 to 36 hours. If the power is out for a longer period, pack the food in dry ice (frozen carbon dioxide). Set the dry ice on heavy cardboard or layers of newspaper, not in direct contact with food. Keep the door closed as much as possible and do not touch the dry ice; it causes frostbite. If food must be removed from the refrigerator or freezer during a repair, place it in the bathtub, layered with newspapers and dry ice.

Before starting any repair, always unplug the refrigerator or disconnect the fuse or circuit breaker. When disconnecting electrical parts, mark the locations of wires and terminals to ensure correct reassembly. Before plugging in the refrigerator, wait an hour. This equalizes the pressure in the sealed cooling system, lessening the start-up strain on the compressor.

A refrigerator may come equipped with an icemaker, or one may be installed later. To make ice, a solenoid-operated water inlet valve at the back of the refrigerator meters in enough water to fill the ice mold. A thermostat switch inside the icemaker determines when the water is frozen and activates the motor and mold heater (if any) to eject the ice into a bin. A shutoff arm, pushed up by accumulating ice, turns off the icemaker when the bin is full. The Troubleshooting Guide for icemakers appears on page 16. The diagram and repairs begin on page 32.

TROUBLESHOOTING GUIDE   continued ►

SYMPTOM	POSSIBLE CAUSE	PROCEDURE
Refrigerator doesn't run and light doesn't work	No power to refrigerator	Check that refrigerator is plugged in; check for blown fuse or tripped circuit breaker *(p. 132)* □○
	Power cord loose or faulty	Test power cord *(p. 133)* ▭●
Refrigerator doesn't run, but light works	Temperature control turned off	Check temperature control
	Temperature control faulty	Test temperature control *(dial type, p. 22; console type, p. 23)* ▭●
	Compressor overheated	Clean condenser coils *(p. 17)* □○
	Condenser fan faulty	Service condenser fan and motor *(p. 29)* ▭●▲
	Overload protector faulty	Test overload protector *(p. 31)* ▭●
	Compressor relay faulty	Test compressor relay *(p. 30)* ▭●▲
	Defrost timer faulty	Test defrost timer *(p. 28)* ▭●▲
	Compressor faulty	Service compressor *(p. 31)* ▭●▲, or call for service
Refrigerator starts and stops rapidly	Condenser coils dirty	Clean condenser coils *(p. 17)* □○
	Condenser fan faulty	Service condenser fan and motor *(p. 29)* ▭●▲
	Compressor faulty	Service compressor *(p. 31)* ▭●▲, or call for service
	Overload protector tripping repeatedly	Have electrician check voltage at outlet
Refrigerator runs constantly	Frost buildup	Defrost refrigerator
	Poor door seal	Check door seal *(p. 18)* □○
	Door gasket damaged	Replace door gasket *(p. 20)* ▭●
	Condenser coils dirty	Clean condenser coils *(p. 17)* □○
	Evaporator plate or coils dirty (models with exposed evaporator)	Wash evaporator plate or coils with warm, soapy water; rinse well □○
	Condenser fan faulty	Service condenser fan and motor *(p. 29)* ▭●▲

DEGREE OF DIFFICULTY: □ Easy ▭ Moderate ■ Complex
ESTIMATED TIME: ○ Less than 1 hour ◗ 1 to 3 hours ● Over 3 hours   ▲ Multitester required

TROUBLESHOOTING GUIDE (continued)

SYMPTOM	POSSIBLE CAUSE	PROCEDURE
Refrigerator not cold enough	Temperature control set too high	Move temperature control to lower setting
	Temperature control faulty	Test temperature control (dial type, p. 22; console type, p.23) ◨●
	Condenser coils dirty	Clean condenser coils (p. 17) □○
	Door doesn't close automatically	Door should swing shut when left open halfway. If not, raise front leveling feet or rollers to tilt refrigerator backward slightly □○
	Poor door seal	Check door seal (p. 18) □○
	Door gasket damaged	Replace door gasket (p. 20) ◨●
	Door switch faulty (some models)	Test door switch (p. 21) □○
	Evaporator fan faulty	Service evaporator fan (p. 25) ◨●
	Evaporator clogged by ice	Defrost refrigerator Test defrost heater (p. 26) ◨●▲ Test defrost limit switch (p. 27) ◨●▲ Test defrost timer (p. 28) ◨●▲
	Refrigerant leaking or contaminated	Call for service
Refrigerator too cold	Temperature control set too low	Move temperature control to higher setting
	Temperature control faulty	Test temperature control (dial type, p. 22; console type, p. 23) ◨●
Refrigerator doesn't defrost automatically	Defrost heater faulty	Test defrost heater (p. 26) ◨●▲
	Defrost limit switch faulty	Test defrost limit switch (p. 27) ◨●▲
	Defrost timer faulty	Test defrost timer (p. 28) ◨●▲
Moisture around refrigerator door or frame	Breaker strips faulty	Inspect breaker strips (p. 24) □●
	Energy-saver switch on or faulty	Reset or test energy-saver switch (p. 23) ◨●
	Internal heater defective	Call for service
Ice in drain pan or water in bottom of refrigerator	Drain tube clogged	Clean drain tube (p. 18) □○
Water on floor around refrigerator	Drain pan damaged or misaligned	Check drain pan (p. 18) □○
	Drain tube clogged	Clean drain tube (p. 18) □○
	Icemaker water inlet valve leaking	Check icemaker water inlet valve (p. 35) ◨●▲
Interior light doesn't work	Bulb loose or burned out	Check bulb (p. 21) □○
	Door switch faulty	Test door switch (p. 21) □○
Refrigerator smells bad	Contents spoiled	Remove spoiled food; wash interior of refrigerator with baking soda and warm water
	Drain pan dirty	Wash drain pan (p. 18) □○
	Insulation absorbing moisture through damaged breaker strips	Remove breaker strips and allow insulation to dry; replace strips if damaged (p. 24) ◨●
Refrigerator noisy	Refrigerator not level	Adjust leveling feet □○
	Drain pan rattling	Reposition drain pan □○
	Compressor mountings loose or hardened	Replace compressor mountings (p. 30) ◨●
	Condenser fan damaged	Inspect condenser fan (p. 29) ◨●▲
	Evaporator fan damaged	Check evaporator fan (p. 25) ◨●▲

ICEMAKERS

SYMPTOM	POSSIBLE CAUSE	PROCEDURE
Icemaker doesn't make ice	No power to refrigerator	Check that refrigerator is plugged in; check for blown fuse or tripped circuit breaker (p. 132) □○
	Water supply to icemaker turned off	Turn on water at valve under sink
	Freezer compartment too warm	Freezer must be 10°F or less; check temperature (p. 22) □○
	ON/OFF switch faulty	Test ON/OFF switch (p. 34) ◨●▲
	Holding switch faulty	Test holding switch (p. 34) ◨●▲
	Water inlet valve switch faulty	Test water inlet valve switch (p. 34) ◨●▲
	Motor faulty	Test motor (p. 35) ◨●▲
	Thermostat faulty	Test thermostat (p. 35) ◨●▲
	Water inlet valve filter clogged	Clean water inlet valve filter (p. 35) □○
	Water inlet valve faulty	Test water inlet valve (p. 35) ◨●▲

DEGREE OF DIFFICULTY: □ Easy ◨ Moderate ■ Complex
ESTIMATED TIME: ○ Less than 1 hour ◐ 1 to 3 hours ● Over 3 hours

▲ Multitester required

SYMPTOM	POSSIBLE CAUSE	PROCEDURE
Icemaker doesn't stop making ice	Shutoff arm out of position	Prop up shutoff arm; if ice stops, check shutoff arm *(p. 33)* ▢●
	ON/OFF switch faulty	Test ON/OFF switch *(p. 34)* ▢●▲
Water on floor	Water inlet valve tubes leaking	Check water inlet valve tube connections *(p. 35)* ▢●
Water overflows from icemaker	Icemaker not level	Adjust refrigerator leveling feet
	Water inlet valve switch faulty	Test water inlet valve switch *(p. 34)* ▢●▲
	Water inlet valve faulty	Test water inlet valve *(p. 35)* ▢●▲
Icemaker doesn't eject ice cubes	Motor faulty	Test motor *(p. 35)* ▢●▲
	Holding switch faulty	Test holding switch *(p. 34)* ▢●▲
	Thermostat faulty	Test thermostat *(p. 35)* ▢●▲
Ice cubes discolored or flecked	Water inlet valve filter dirty	Clean water inlet valve filter *(p. 35)* ▢○
	Ice mold worn	Call for service
	Hard water	Have in-line water filter installed; call for service
Ice cubes smell or taste bad	Refrigerator or freezer dirty	Clean refrigerator and freezer; flush drain *(p. 18)* ▢○
	Food spoiled or improperly stored	Remove spoiled food; rewrap food
	Ice bin dirty	Wash ice bin
	Ice old	Discard ice; wash ice bin
	Hard water	Have in-line water filter installed; call for service

DEGREE OF DIFFICULTY: ▢ **Easy** ▢ **Moderate** ■ **Complex**
ESTIMATED TIME: ○ **Less than 1 hour** ● **1 to 3 hours** ● **Over 3 hours**   ▲ **Multitester required**

CLEANING THE CONDENSER COILS

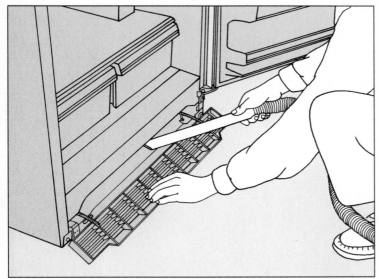

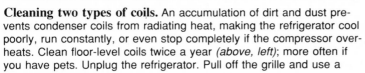

Cleaning two types of coils. An accumulation of dirt and dust prevents condenser coils from radiating heat, making the refrigerator cool poorly, run constantly, or even stop completely if the compressor overheats. Clean floor-level coils twice a year *(above, left)*; more often if you have pets. Unplug the refrigerator. Pull off the grille and use a vacuum cleaner with a wand attachment to remove dust and pet hair that accumulate behind the grille. Clean rear-mounted coils *(above, right)* yearly, using a stiff brush or a vacuum cleaner with a brush attachment. If the coils are greasy, wash them with warm, soapy water, taking care not to drip water on other parts of the refrigerator.

CLEANING THE DRAIN AND DRAIN PAN

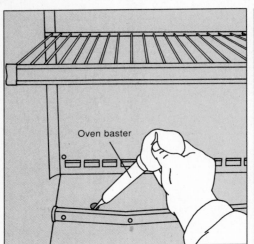
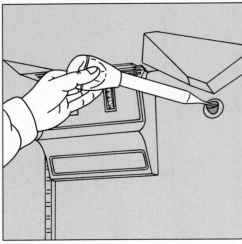

Oven baster

Cleaning the drain and drain pan.
The drain opening may be in the floor of the refrigerator under the storage drawer *(far left)*, or at the top of the back wall, behind a pull-off drain trough *(near left)*. To clean the drain, use an oven baster to force a solution of hot water and baking soda or bleach into the opening. To clear a stubborn clog, insert a length of 1/4-inch round plastic tubing into the drain and push it through to the drain pan below, then pull it out. The drain pan is located under the refrigerator, behind the front grille. Wash the pan regularly with a warm baking soda solution. If the pan rattles, it may be located too close to the compressor; reposition it.

CHECKING THE DOOR SEAL

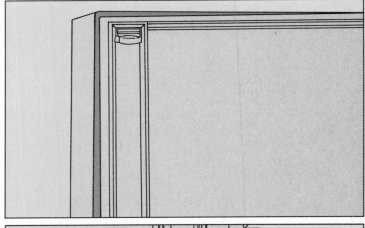

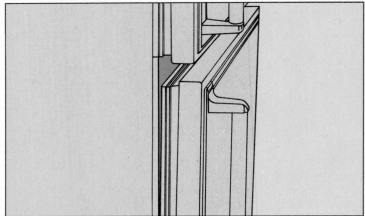

Checking the door seal. Open the door and examine all four sides of the door gasket for tears. Feel the gasket for brittleness or cracks. If the gasket shows damage, replace it *(page 20)*. If not, close the door and check the seal between gasket and cabinet for obvious gaps. Next open the door and shut it on a dollar bill, as shown. Slowly pull the dollar bill out of the door. If the gasket seals properly, you will feel tension as it grips the bill. Repeat this test all around the door.

For the door to close tightly and automatically, the refrigerator should tilt backward slightly. If the door seal isn't tight, have a helper push the refrigerator backward while you adjust the feet or rollers to raise the front of the refrigerator slightly. If the gasket still does not seal tightly, check the door for sagging or warping *(next step)*.

Checking the door for sagging or warping. If the refrigerator or freezer door sags on its hinges *(above, top)*, a poor seal will result. First adjust the hinges *(page 19)*. If the door gasket does not press flat against the cabinet, or the door appears warped *(above, bottom)*, the inner and outer door panels may be out of alignment. Try adjusting the hinges; if this doesn't work, reposition the door panels *(page 19)*.

CHECKING THE DOOR SEAL (continued)

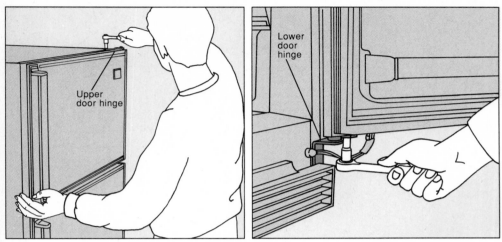

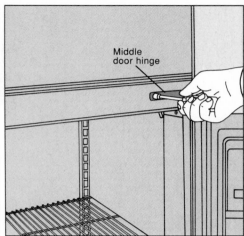

Adjusting the door hinges. Unplug the refrigerator. To correct a sag in an upper freezer door or in a full-length refrigerator door, use a socket wrench or hex wrench to loosen the bolts on the upper hinge *(above, left)*. Slight adjustments may also be made to a refrigerator door by loosening the bolts on the lower hinge *(above, center)*. Lift or push the door square with the refrigerator cabinet and retighten the hinge bolts. To correct a sag in a lower refrigerator door, open the door and loosen the screws on the middle hinge *(above, right)*. Shift the hinge slightly toward the outside of the cabinet and tighten the screws.

To straighten a warped door, loosen the hinge bolts nearest the warped area, push the door tight to the cabinet and retighten the bolts. Check that the door rests flat against the cabinet all around the gasket. If not, the door panels must be realigned *(next step)*.

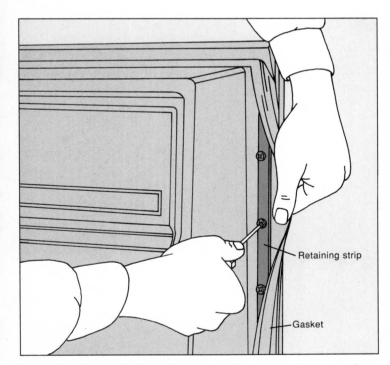

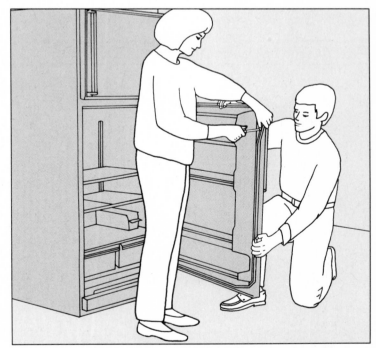

Loosening the retaining strip screws. Unplug the refrigerator. Open the door and pull back the door gasket to expose the metal or plastic retaining strips. Loosen, but do not remove, all the screws along the strip *(above)*.

Aligning the inner and outer door panels. Grasp the outer door panel at the top and side and twist it opposite to the warp, flattening the door. While you hold the door in this position, have a helper partially retighten the retaining screws *(above)*. Close the door and check that the warp has been corrected. If so, open the door and hold it while your helper tightens the screws securely. If the door is still warped, try adjusting it again. If the warp persists, the door may need to be replaced *(page 20)*.

REPLACING A DOOR GASKET

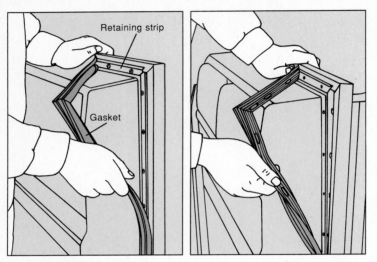

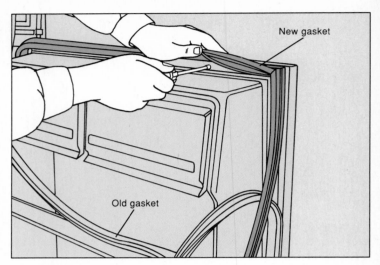

1 **Removing the gasket.** Before removing the old gasket, soak the new one in warm water to soften it and make installation easier. Unplug the refrigerator, pull back the gasket to expose the retaining strip and loosen the screws *(page 19)*. Try to pull out the gasket from behind the retaining strip. On newer models, it will come free *(above, left)*; on older models, the screws pass through holes in the gasket itself *(above, right)*. To free such a gasket, remove the screws from the retaining strip along the top edge of the door and one-third of the way down each side. Pull the upper part of the gasket away from the retaining strip.

2 **Installing a new gasket.** On a newer model, start at an upper corner and simply insert the rear flange of the gasket behind the retaining strip, partially tightening the screws as you go. To keep the door from warping, have a helper hold it at the top and side as you tighten the screws *(page 19)*. When the gasket is installed all around the door, tighten the screws securely.

On an older model, push the flange of the new gasket behind the retaining strip. Install and partially tighten the screws *(above)*. Then remove the screws from the lower part of the retaining strip and pull out the rest of the old gasket. Push the rest of the new gasket into place. Starting at the bottom corners, reinstall and partially tighten the screws, then tighten all the screws securely.

INSTALLING UPPER AND LOWER DOORS

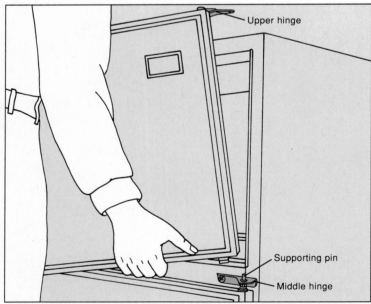

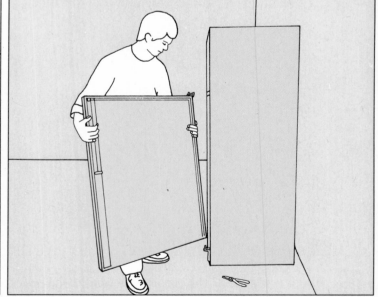

Removing and replacing the doors. Unplug the refrigerator. Use a socket wrench or hex wrench to remove the bolts holding the upper hinge to the refrigerator cabinet and lift the upper door off the supporting pin *(above, left)*; save any hinge washers. To install a new door, remove the upper hinge from the old door and insert it in the new door, replacing any washers. Rest the door on the lower supporting pin and, aligning the door with the cabinet, install the upper hinge. To remove a lower door, you must first remove the upper door. Next, unscrew the middle hinge from the refrigerator cabinet and lift the door off the lower supporting pin *(above, right)*. Install a new door as described, replacing any hinge washers.

CHECKING THE LIGHT BULB AND DOOR SWITCH

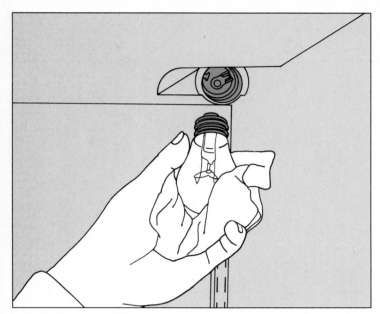

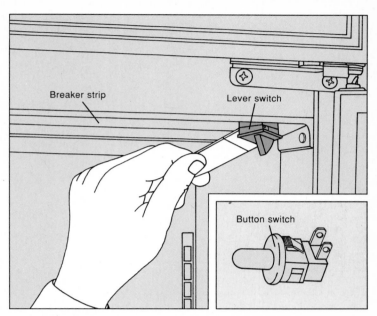

1 **Checking the light bulb.** In most refrigerators, a single door switch controls the interior light. In others, the same switch or a second door switch controls an evaporator fan as well. To test the evaporator fan switch only, go to step 2. If the interior light doesn't glow when you open the door, first check for a burned-out bulb. Protecting your hand with a rag or glove, unscrew the bulb, as shown. Replace the bulb with a new one of the same wattage. If it does not light, the door switch may be faulty; remove and test it *(step 2)*. If you suspect that the light stays on when the refrigerator door is closed, warming the interior, press the door switch by hand. If the light stays on, remove and test the switch.

2 **Removing the door switch.** Unplug the refrigerator. Pry out a lever door switch *(above)* or a button switch *(inset)* using a putty knife, the blade padded with masking tape to prevent scratching the breaker strip. On some older refrigerators, the breaker strip must first be removed to free the switch from behind it *(page 24)*.

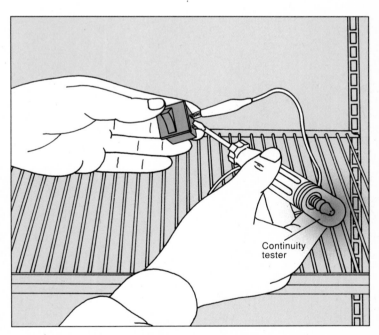

3 **Disconnecting the door switch.** Ease the switch housing from the breaker strip and pull it out to expose a few inches of wiring. The switch will have two or four terminals. Remove the push-on connectors, labeling the wire positions for reassembly. If the wires are burned or corroded, splice on new ones *(page 136)*.

4 **Testing the door switch.** Place a continuity tester probe on each terminal of a two-terminal switch. For a light switch, when the switch button is out the continuity tester should light *(above)*; when the button is depressed the continuity tester should not light. An evaporator fan switch will give the opposite result.

   If the switch has four terminals, it is a combination evaporator fan and light switch. There should be continuity between one pair of terminals with the switch button out. Depress the switch button, and the other pair of terminals should show continuity. If the switch fails any test, replace it. Connect the new switch to the wire leads and snap the switch into the breaker strip.

TESTING AND REPLACING THE TEMPERATURE CONTROL (Dial type)

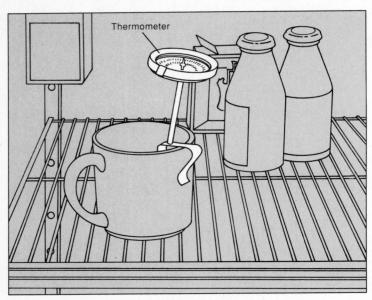

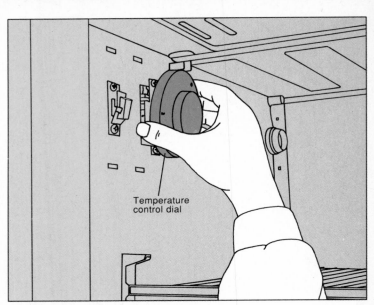

1 **Testing the refrigerator and freezer temperature.** The ideal temperature for the refrigerator is between 38°F and 40°F; for the freezer compartment, between 0°F and 8°F. (The freezer temperature may be about 10°F higher in a single-door refrigerator.) To test the temperature of the refrigerator compartment, place a cup of water in it for 24 hours. (In a freezer, use cooking oil.) Place a cooking thermometer in the liquid for three minutes. If the temperature is too cold or too warm, adjust the temperature control. If the problem persists, test the control (dial type, *step 2*; console type, *page 23*).

2 **Removing the dial.** Unplug the refrigerator. Turn the dial to its coldest setting. If the dial has a screw in the center, unscrew it, then pull the dial straight off its shaft, as shown.

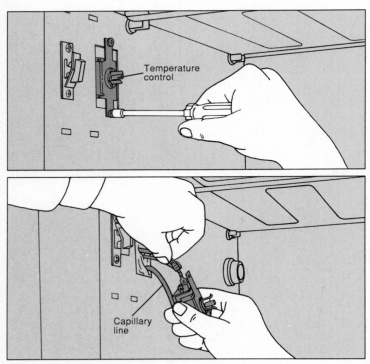

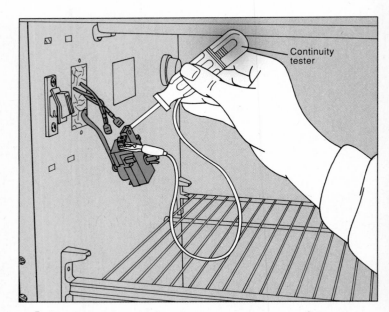

3 **Removing the temperature control.** Remove the screws securing the temperature control to the wall of the refrigerator *(above, top)*. Pull out the control to expose a few inches of electrical wiring, taking care not to bend or damage the metal capillary line. Pull off the wire connectors and the ground wire *(above, bottom)*.

4 **Testing and replacing the temperature control.** Touch a continuity tester probe to each terminal *(above)*. With the control at its coldest setting *(step 2)*, the tester should light, indicating a closed circuit. Turn off the control by twisting the shaft in the opposite direction until it stops, then retest; the tester should not light. To install a new temperature control, pull the capillary line of the old control out of its opening in the refrigerator wall. Set the new control to its coldest setting and carefully thread the capillary line into the opening without kinking it. Attach the wires to the terminals, screw the control in place on the wall, and reattach the dial.

TESTING AND REPLACING THE TEMPERATURE CONTROL (Console type)

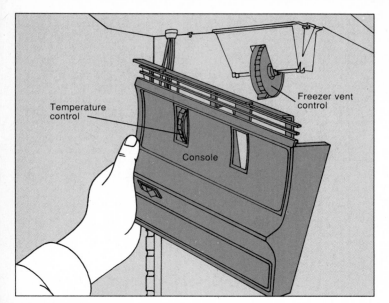

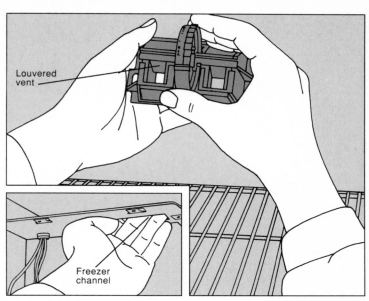

1 **Removing the temperature control console.** Test and adjust the temperature of the refrigerator or freezer *(page 22)*. If the temperature is not within the acceptable range, unplug the refrigerator. Unscrew the console and carefully remove it from the refrigerator wall. The temperature control (and energy-saver switch, if any) is mounted on the console; let it dangle by its wiring. The freezer vent control will remain attached to the refrigerator wall.

2 **Checking the freezer control.** Unscrew the freezer vent control from the refrigerator wall. Use a hair dryer set on LOW to melt any ice blocking the louvered vent. Remove any food that might have fallen into the vent from the freezer. Finally, reach into the freezer channel in the refrigerator wall with your fingers *(inset)* and check for obstructions. Reinstall the freezer vent control.

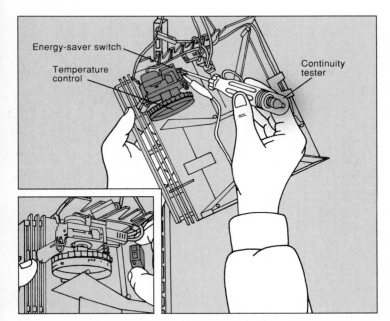

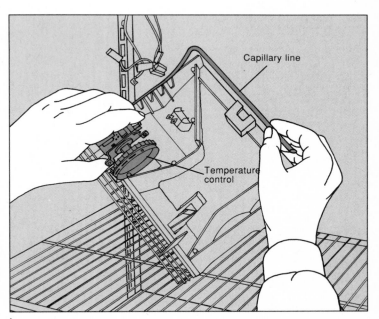

3 **Testing the energy-saver switch and temperature control.** If the console has an energy-saver switch, remove the push-on connectors *(inset)* and test the switch for continuity. With a probe on each terminal, the tester should light when the switch is on. If the switch is faulty, pull it off the console, snap on a new switch and reattach the push-on connectors.

To test the temperature control, remove the push-on connectors and the ground wire from its terminals. Touch a tester probe to each terminal *(above)*. With the control turned to its coldest setting, the tester should light; with the control turned off, it should not light. Replace a faulty control.

4 **Replacing the temperature control.** Note the position of the temperature control's capillary line in the console; the new one must be installed the same way. Pull out the old control and snap in a new one, threading the capillary line into place in the console. Remount the console on the refrigerator wall.

REPLACING THE BREAKER STRIPS

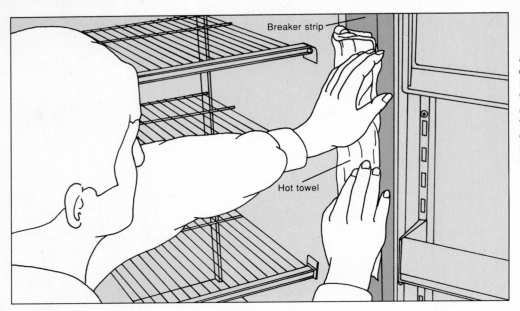

Breaker strip

Hot towel

1 **Softening the breaker strip.** Damaged breaker strips allow moisture to enter the insulation between the inner and outer walls of the refrigerator or freezer, causing odor and reducing cooling efficiency. Inspect the breaker strips around the inner frame for warps or cracks. To replace a damaged breaker strip, first unplug the refrigerator. Because a breaker strip is brittle when cold, soften the strip before attempting removal by pressing a hot, wet towel against it along its entire length *(left)*.

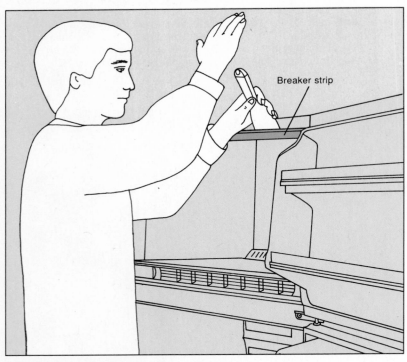

Breaker strip

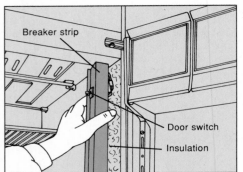

Breaker strip

Door switch

Insulation

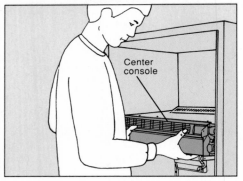

Center console

2 **Removing and replacing a breaker strip.** The way breaker strips are installed varies from model to model. Before beginning work, check to see how the strips are attached to each other and to the refrigerator cabinet. Most can be pried from the cabinet with a wide putty knife *(above)*. Be careful not to damage the foam insulation behind the breaker strip. If the insulation is damp or ice-clogged, or smells bad, leave it uncovered for a few hours or dry it with a hair dryer set on LOW. To replace the breaker strip, simply snap a new one in place.

Other kinds of breaker strips are more complicated to remove. On some models, you must first remove the unbroken strips to release the broken one. In other cases, the breaker strip may have a door switch mounted in it. Partially free the strip *(right, top)* and disconnect the wires from the back of the switch before removing the strip completely. Transfer the switch to the new breaker strip before installing it.

On some refrigerators, the top and bottom breaker strips are attached to the side strips with sealant. Cut through the sealant with utility knife before removing the damaged breaker strip. After replacing the strip, reseal the corners with an arsenic-free sealant rated for use in food compartments.

The most difficult breakers to remove and replace are those that run from the bottom of the refrigerator to the top of the freezer and are held in place by the center console. To free these strips, first remove the screws that hold the console in place, then pull the console forward *(right, bottom)* and rest it in the freezer compartment without disconnecting the wires. Replace the breaker strip as described above and reinstall the console.

TESTING AND REPLACING THE EVAPORATOR FAN

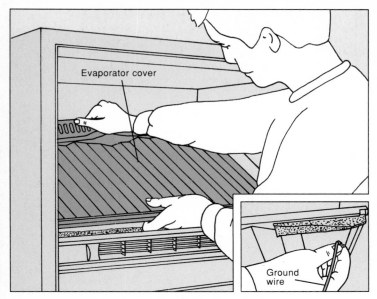

Evaporator cover

Ground wire

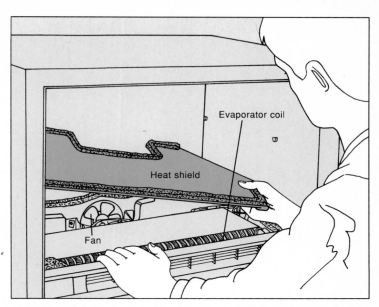

Evaporator coil

Heat shield

Fan

1 **Removing the evaporator cover.** The evaporator coils are part of the sealed refrigeration system and should only be serviced professionally. But the evaporator fan, the defrost heater and the defrost limit switch—all located under the evaporator cover—are easily tested and replaced, and are more likely to cause problems. On most models, the evaporator cover is also the freezer bottom. To remove it, first unplug the refrigerator. Remove the screws around the edges of the cover and lift it partway out. Unclip the ground wire from its underside *(inset)* and remove the cover from the freezer, as shown. If the freezer has an icemaker, remove it first *(page 32)*. On side-by-side models, the evaporator cover may be located at the top of the refrigerator compartment or at the back of the freezer.

2 **Removing the insulation and heat shield.** Carefully peel off the tape that holds the rigid sheet of foam insulation in place and remove it. Lift the metal heat shield from the evaporator coil compartment *(above)*. With a hair dryer set on LOW, melt any ice that has built up around the fan blades, taking care not to melt plastic components. Remove any objects that have fallen into the compartment through the air vent.

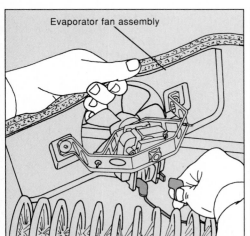

Evaporator fan assembly

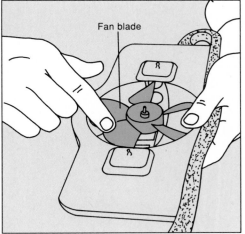

Fan blade

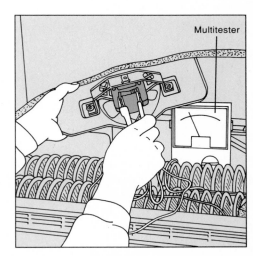

Multitester

3 **Checking the evaporator fan.** If the fan housing is attached to the cabinet with screws, remove them. Lift the fan assembly a few inches and disconnect the push-on connectors and the ground wire *(above, left)*, using long-nose pliers if they do not pull off easily. If the fan blade is damaged, replace it. Unscrew the nut at the center of the fan blade, pull the blade off the motor shaft and slide on a new blade, replacing any washers. Hold the fan horizontally and spin the blade to check for binding in the motor *(above, right)*. If the blade does not spin freely, replace the motor. Remove the fan blade, unscrew the smaller bracket at the front of the motor and remove the motor from the housing. Install the new motor and replace the bracket and the fan blade, taking care not to reverse the blade.

4 **Testing the evaporator fan motor.** Set a multitester at RX1 and touch a probe to each motor terminal. The meter should show some resistance. If not, install a new motor *(step 3)*, and reinstall the fan. Test the defrost heater *(page 26)* before replacing the evaporator cover.

TESTING AND REPLACING THE DEFROST HEATER

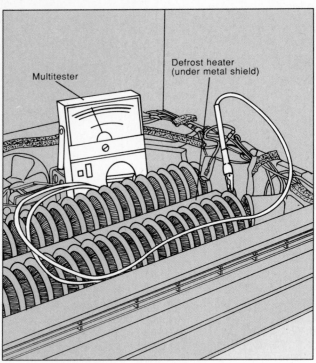

Multitester

Defrost heater (under metal shield)

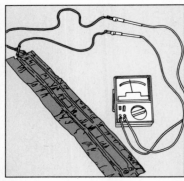

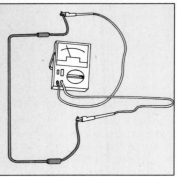

1 **Testing the defrost heater.** The defrost heater element may be enclosed within a glass tube and hidden beneath a metal reflector shield between the evaporator coils *(far left)*. Alternatively, it may be wrapped in aluminum foil *(near left, top)*, or it may be an exposed metal rod *(near left, bottom)*. All elements are tested the same way. Unplug the refrigerator and remove the evaporator cover, insulation and heat shield *(page 25)*. Pull the wire connectors from the terminals at each end of the defrost heater. Set a multitester at RX1 and attach a probe to each terminal. The meter should show medium to high resistance. If not, replace the element. If the element is good, test the defrost limit switch *(page 27)*.

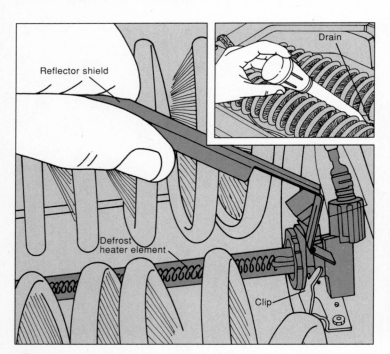

Reflector shield

Drain

Defrost heater element

Clip

2 **Removing the defrost heater.** Unhook the element's reflector shield from the clips at each end, and carefully lift the element out of its brackets. This will expose the opening to the drain tube. Before installing a new element, clean the drain tube by using an oven baster to force a solution of hot water and baking soda or bleach into the opening *(inset)*.

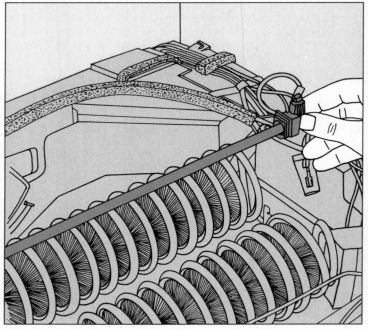

3 **Installing a new defrost heater.** Do not touch the glass surface of the new heater element; oils from your skin will cause hot spots. If you do touch the element, wipe it thoroughly with a paper towel. Plug the new element in the same position as the old one and replace all clips or fasteners. Next, reconnect the push-on connectors to the element terminals and snap in the reflector shield, if any. If you are not testing the defrost limit switch *(page 27)*, reinstall the evaporator cover, insulation and heat shield *(page 25)*. Take special care to reconnect the ground wire.

TESTING AND REPLACING THE DEFROST LIMIT SWITCH

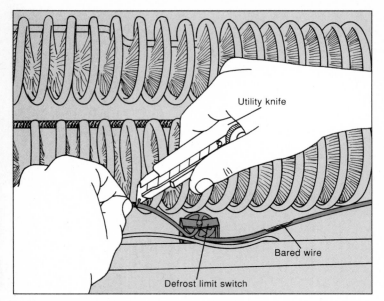

Utility knife

Bared wire

Defrost limit switch

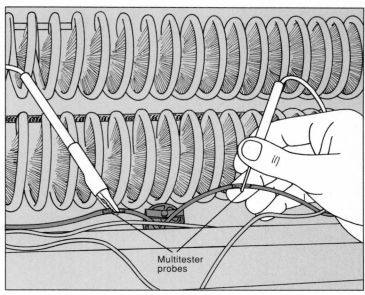

Multitester probes

1 **Baring the defrost limit switch wires.** The defrost limit switch, located next to the evaporator, turns off the defrost heater when all the frost has melted during the defrost cycle. To reach it, unplug the refrigerator and remove the evaporator cover, insulation, and heat shield *(page 25)*. If the defrost limit switch has push-on connectors, disconnect them. On most refrigerators, however, the switch is permanently wired. To test such a switch, you must bare the two switch wires. Taking care not to cut or fray the copper strands inside, use a sharp utility knife to remove a small patch of the plastic insulation around each wire *(above)*.

2 **Testing the defrost limit switch.** The switch should have complete continuity when it is cold, and no continuity when it is warm. To test the switch, place a plastic bag full of dry ice on it for 20 minutes. Remove the ice and touch a multitester probe to each of the two exposed wires (or to each switch terminal). The tester should show continuity. Then warm the switch using a hair dryer set on LOW. The multitester needle should swing downscale to show resistance. If the switch fails either test, replace it. If the switch is good, tape the exposed wires *(step 3)*.

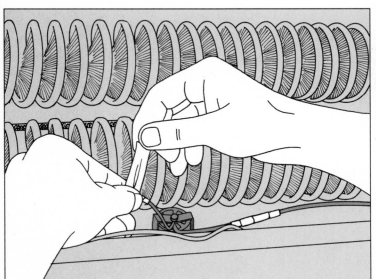

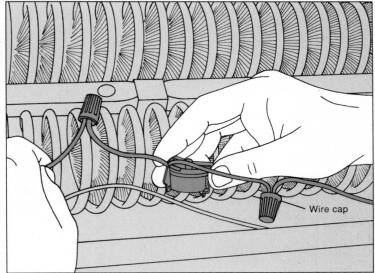

Wire cap

3 **Taping wires or replacing the defrost limit switch.** If the switch is good, reseal the bared wires with electrical tape, covering the cut patches and one inch on each side *(above, left)*. If the defrost limit switch has terminals, reconnect the push-on connectors. If the switch is faulty, cut the wires at the bared spots. Splice them to the wires of a new switch *(page 136)* using wire caps *(above, right)* or crimp connectors waterproofed with a dab of silicone sealant. Snap the new switch into place. Reinstall the evaporator cover *(page 25)*. Take special care to reconnect the ground wire.

TESTING AND REPLACING THE DEFROST TIMER

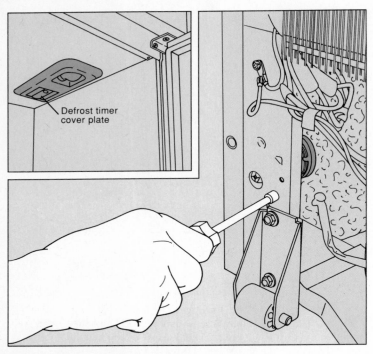

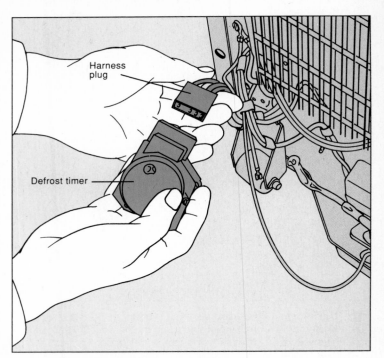

1 Removing the defrost timer. The defrost timer is usually located in the compressor compartment at the back of the refrigerator, as shown, but it may also be found behind the front grille, in the thermostat control console, or behind a cover plate inside the refrigerator *(inset)*. To remove the defrost timer, unplug the refrigerator and unscrew the timer from the cabinet *(above)*.

2 Disconnecting the wires. Disconnect the green ground wire from the timer. The timer is linked to the wiring by a harness plug, which houses four connections. To help you reconnect it in the proper position, mark one side of both the plug and the defrost timer with masking tape before pulling apart the plug.

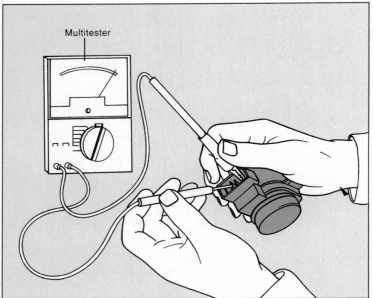

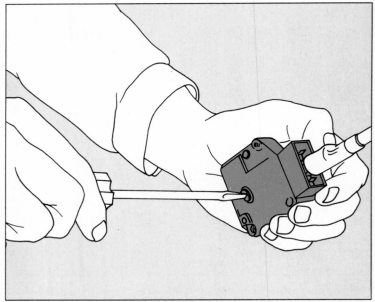

3 Testing the defrost timer. Find the common terminal of the timer, usually connected to the white wire of the harness plug (if the terminals are numbered, it is number 3). If you can't identify it this way, consult the wiring diagram *(page 138)*. Attach one multitester probe to the common terminal and, with the meter set at RX100, touch the other probe to each of the other three terminals *(above)*. Two of these pairs should have full continuity while the third should have no continuity.

Then, using a screwdriver, turn the defrost timer switch manually *(above)* until you hear a click. Test the timer again the same way. Two of the three terminal pairs should show continuity, while the third—not the same one as before—should not. In either test, if all three pairs have continuity, or if only one does, the defrost timer is faulty. To install a new defrost timer or to reinstall the old one, reconnect the green ground wire and then reconnect the harness plug. Screw the defrost timer to the refrigerator cabinet and replace the back panel, if any.

SERVICING THE CONDENSER FAN

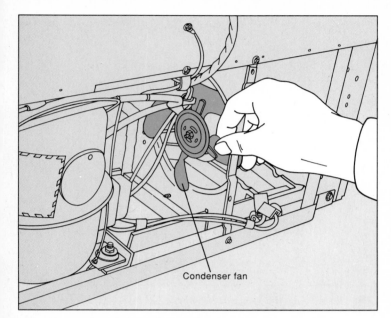

Condenser fan

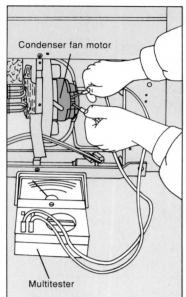

Condenser fan motor

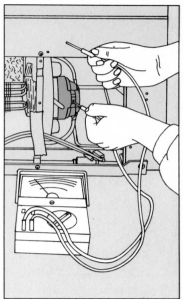

Multitester

1 **Inspecting the condenser fan.** On refrigerators with the condenser at floor level, the condenser fan circulates cool air over the condenser coils. Unplug the refrigerator and pull it away from the wall. Clean the fan blade, and turn it to see if the blade rotates freely *(above)*. If the motor binds, test it *(step 2)*. If the fan blade is damaged, unscrew the nut that holds it to the motor shaft and pull it off. Install a new fan blade, replacing any washers, and tighten the nut.

2 **Testing the condenser fan motor.** Disconnect the wires to the fan motor. Set a multitester at RX10 and touch one probe to each terminal *(above, left)*. The multitester needle should move to the medium range of the scale, showing partial resistance; a low reading means the motor is faulty. Next set the multitester at RX1000 and touch one probe to the motor terminals and the other to any unpainted metal part of the refrigerator *(above, right)*. If the multitester needle moves, the motor is grounded and should be replaced.

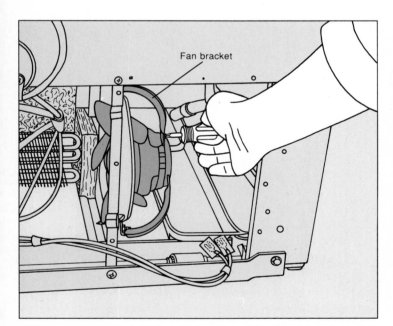

Fan bracket

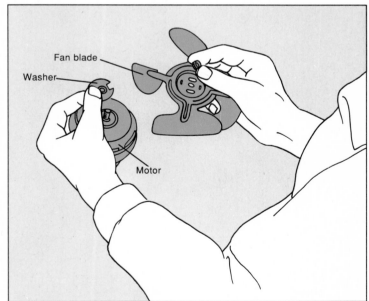

Fan blade

Washer

Motor

3 **Removing the motor.** Unscrew the brackets that hold the fan motor to its housing; if necessary, unscrew the mounting plate from the motor. Slide the motor backward out of the housing.

4 **Replacing the motor.** Remove the fan blade from the old motor and attach it to the new motor *(step 1)*, replacing any washers. Install the new motor in its housing by screwing the brackets in place. Reattach the wires to the motor terminals and reconnect the green ground wire.

SERVICING THE COMPRESSOR

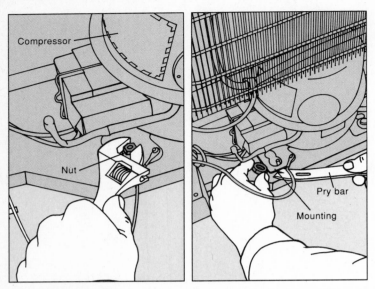

Compressor

Nut

Pry bar

Mounting

1 **Replacing the compressor mountings.** Unplug the refrigerator and pull it away from the wall. If it has an access panel, remove it. Using an adjustable wrench or socket wrench, unscrew the nut securing one foot of the compressor *(above, left)*. Jack up the compressor foot with a pry bar, just far enough to pull out the shock absorber mounting from beneath it *(above, right)*. Slip a new mounting in place and lower the compressor foot. Replace the washer and tighten the nut. Repeat for each mounting; do not remove more than one at a time.

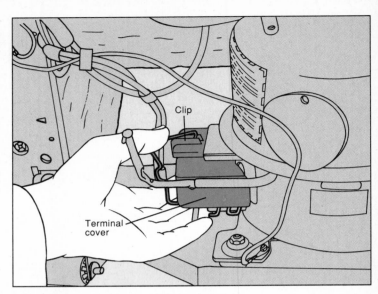

Clip

Terminal cover

2 **Removing the compressor terminal cover.** The compressor is part of the closed refrigeration system and should be replaced only by a professional service technician. You can, however, test the compressor and certain components. Unplug the refrigerator, pull it away from the wall and remove the rear access panel, if any. A small box mounted on the compressor cover protects the relay, overload protector and capacitor, if any. Release the wire retaining clip that holds the cover in place *(above)* and slip off the cover and the clip. **Caution**: If the compressor has a capacitor, discharge it before proceeding *(page 111)*.

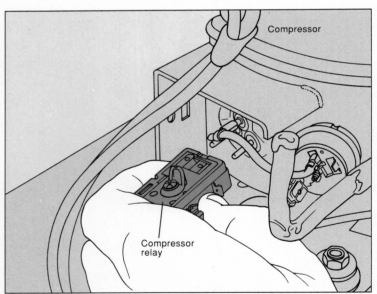

Compressor

Compressor relay

3 **Removing and testing the compressor relay.** Pull the relay straight off the compressor without twisting it *(above)*. If the relay has an external wire coil, hold the relay so that the word TOP is up. Set a multitester at RX1 and place the probes on the terminals marked S and M. The multitester needle should not move. Next remove the probe from M and place it on the side terminal marked L. Once again, the needle should not move. Finally, remove the probe from S and place it on M. The needle should sweep across the scale, showing full continuity.

Now turn the relay upside down (you'll hear a click as the magnetic switch inside the relay engages) and perform the same tests *(above,*

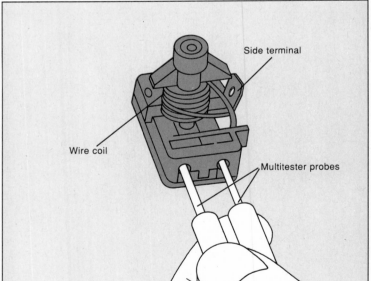

Side terminal

Wire coil

Multitester probes

right). You should get the opposite results: continuity between terminals S and M and between S and L; no continuity between M and L. If the relay fails any of these tests, replace it: Push the new relay onto the compressor terminals and replace the terminal cover. If the relay passes these tests, test the overload protector.

If the relay has no wire coil, hold the relay so that the word TOP is up. (Otherwise, hold it either way). Set a multitester at RX1 and touch a probe to each terminal. The needle should sweep to the middle of the scale. A high or low reading means the relay has failed; replace it as above. If the relay passes the tests, test the overload protector *(step 4)*.

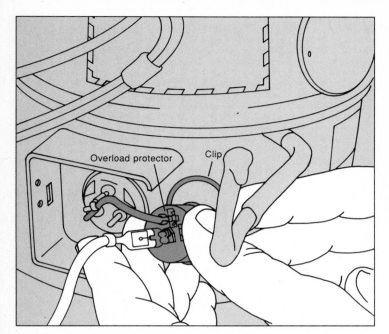

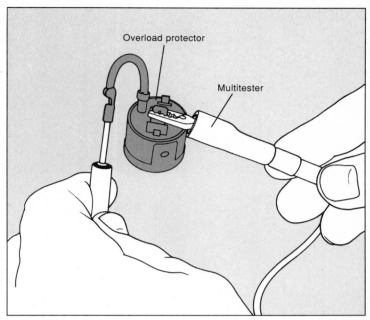

4 **Removing the overload protector.** Using a screwdriver, gently pry open the circular spring clip that secures the overload protector to the compressor and snap out the protector, as shown. Pull the two wire connectors off their terminals.

5 **Testing the overload protector.** Set a multitester at RX1 and touch a probe to each overload protector terminal. The multitester needle should sweep across the scale, showing full continuity. If the overload protector passes this test, test the compressor *(step 6)*. If not, replace the overload protector. Reattach the push-on connectors to the new overload protector, clip it in place on the compressor and replace the terminal cover.

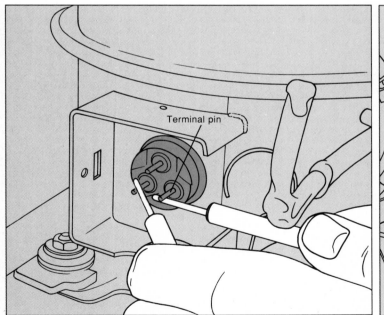

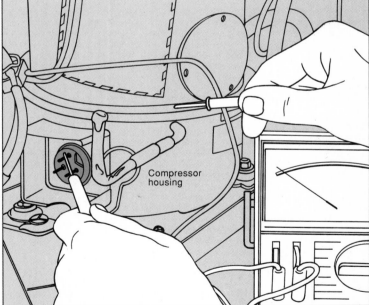

6 **Testing the compressor.** With a multitester set at RX1, test each of the three terminal pins against each of the other two *(above, left)*. Each pair should show continuity. Then, with the multitester set at RX1000, place one probe against the metal housing of the compressor *(above, right)*; if necessary, scrape off a little paint to ensure contact with bare metal. Place the other probe on each of the three compressor terminals in turn. If any of the three terminals shows continuity with the housing, the compressor is grounded. If the compressor fails either test, call for service. If it passes the tests, reinstall the overload protector, relay, terminal cover and rear panel, if any.

ICEMAKERS

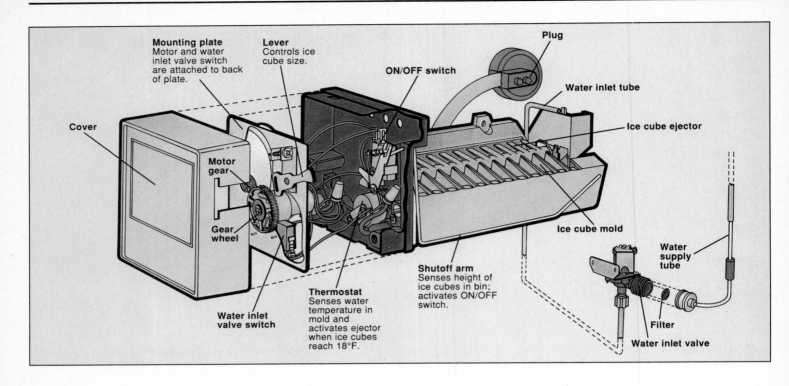

Mounting plate Motor and water inlet valve switch are attached to back of plate.

Lever Controls ice cube size.

ON/OFF switch

Plug

Water inlet tube

Ice cube ejector

Cover

Motor gear

Gear wheel

Ice cube mold

Water inlet valve switch

Thermostat Senses water temperature in mold and activates ejector when ice cubes reach 18°F.

Shutoff arm Senses height of ice cubes in bin; activates ON/OFF switch.

Water supply tube

Filter

Water inlet valve

ACCESS TO THE ICEMAKER

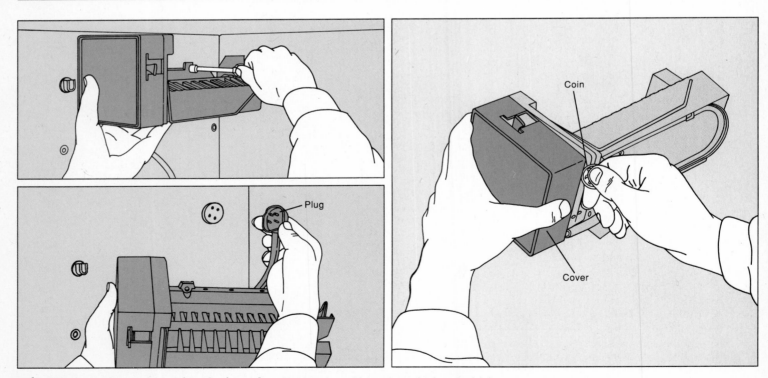

Plug

Coin

Cover

1 **Disconnecting and opening the icemaker.** Unplug the refrigerator and take out the ice bin. Remove the screw from the icemaker's bottom bracket and, supporting the icemaker with one hand, remove the top screws and clips *(left, top)*. Lower the icemaker and unplug it from the refrigerator wall *(left, bottom)*. To remove the cover, insert the edge of a coin into one of the slots at the bottom and twist it *(above, right)*.

ACCESS TO THE ICEMAKER (continued)

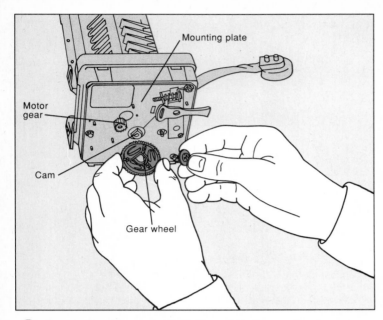

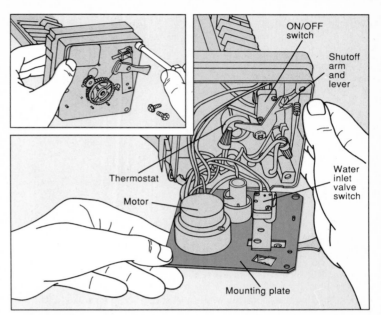

2 **Inspecting the gear wheel.** Turned by the small motor gear, the gear wheel controls the ice cube ejector and the switches. Before removing the mounting plate *(step 3)*, inspect the gear wheel for damage. To replace a damaged wheel, remove the screw at its center and pull the gear off its cam; save the washer. Place a new gear wheel on the cam and replace the washer and screw.

3 **Removing the mounting plate.** If the icemaker doesn't have a large gear wheel outside the mounting plate *(previous step)*, do not remove the plate; call for service. If there is a gear wheel, remove the screws around the edges of the mounting plate *(inset)* and pull it away from the icemaker. You now have access to the motor, switches, thermostat and shutoff arm *(above)*.

REPLACING THE ICEMAKER SHUTOFF ARM

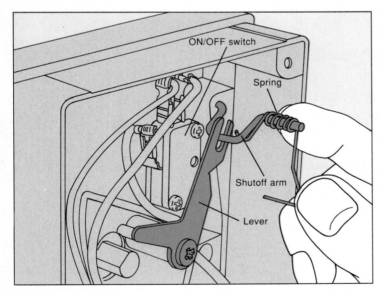

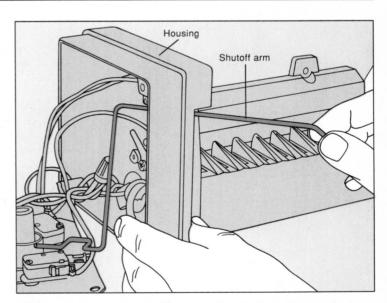

1 **Releasing the spring.** When the shutoff arm is pushed up by accumulating ice, it shifts a lever that turns off the icemaker. To inspect the arm and lever, unplug the refrigerator, take out the icemaker *(page 32)* and remove the mounting plate. Check that the spring is engaged on the shutoff arm *(above)*, and that the arm fits into the lever slot. To remove the arm, first disengage the spring carefully.

2 **Removing the shutoff arm.** Disengage the arm from the lever. Push the arm forward, turning it when necessary to work it out through its hole in the housing *(above)*. To install a new shutoff arm, slide it in through the front of the housing. Engage the arm in the lever and replace the spring. Replace the mounting plate and cover, and reinstall the icemaker.

TESTING ICEMAKER SWITCHES

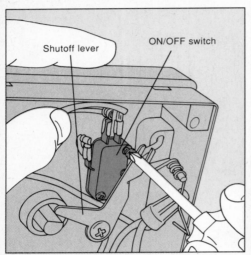

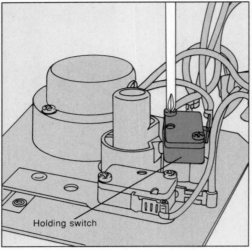

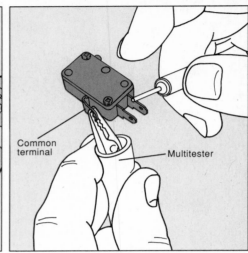

Testing the ON/OFF switch and the holding switch. Although they are identical, these switches perform different functions. The ON/OFF switch turns off the icemaker when the shutoff arm is raised, and the holding switch keeps the power going while the ejector pushes out the ice cubes. To reach the switches, unplug the refrigerator, take out the icemaker *(page 32)* and remove the mounting plate *(page 33)*. The ON/OFF switch is located behind the shutoff lever *(above, left)*; the holding switch is mounted next to the motor *(above, center)*. Label and disconnect the wires, and test both switches the same way: Clip one probe of a multitester set at RX1—or a continuity tester—to the common terminal on the side of the switch. Touch the other probe to each of the other two terminals in turn. With the switch button out *(above, right)*, the tester should show continuity through one terminal and resistance through the other. Press the button in; the results should reverse. If not, replace the switch. Screw a new switch in place, reconnect the wires, and reassemble and reinstall the icemaker.

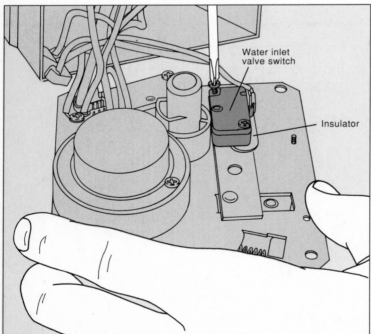

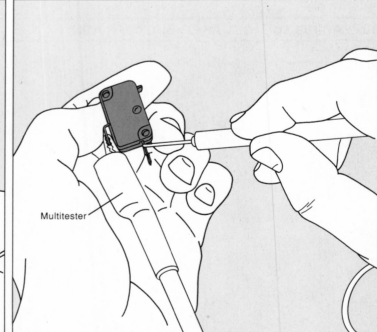

Testing the water inlet valve switch. This switch turns on the water inlet valve, allowing it to meter water into the ice mold. To test the switch, unplug the refrigerator, take out the icemaker *(page 32)* and remove the mounting plate *(page 33)*. Unscrew the switch from the plate *(above, left)*; save the card-like insulator underneath it. Disconnect and label the wires. With a multitester set at RX1—or a continuity tester—touch one probe to each switch terminal *(above, right)*. With the button out, the tester should show continuity; with the button in, resistance. If the switch fails the test, replace it. Position the insulator on the mounting plate with its slanted edge toward the center. Put the switch on top of it and screw it in place. Reconnect the wires, and reassemble and reinstall the icemaker.

TESTING THE ICEMAKER MOTOR AND THERMOSTAT

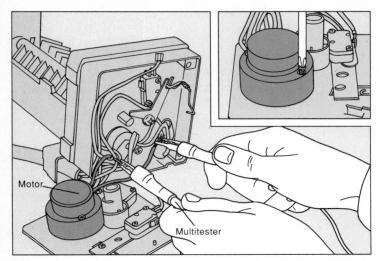

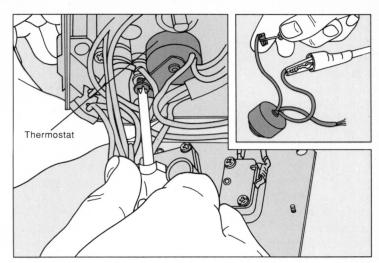

1 **Testing the motor.** Unplug the refrigerator, take out the icemaker *(page 32)* and remove the mounting plate *(page 33)*. Disconnect the motor wires: One wire has two connectors that pull off the switch terminals; the other is connected with a wire cap. Set a multitester at RX10 and touch a probe to each wire *(above)*. The tester should indicate 400 to 600 ohms of resistance. If not, unscrew the motor from the mounting plate *(inset)* and install a new motor, making sure the small gear on the motor meshes with the large gear on the mounting plate. Reconnect the wires, and reassemble and reinstall the icemaker. If the motor tests OK, next test the thermostat.

2 **Testing the thermostat.** Unscrew the clamp *(above)* to free the thermostat. Label and disconnect the wires. If the thermostat has three wires *(inset)*, clip one probe of a multitester set at RX1—or a continuity tester—to the shorter of the two wires with connectors. Touch the other probe to the other wires in turn. A warm thermostat should show continuity through one wire and resistance through the other. Put the thermostat in a freezer at 10°F or less for 15 minutes and test again; the results should reverse. (If the thermostat has two wires, they should show resistance when warm and continuity when cold.) To install a new thermostat, put a dab of factory-specified metallic putty on its back and stick it in place. Screw on the clamp, reconnect the wires, and reassemble and reinstall the icemaker.

SERVICING THE ICEMAKER WATER INLET VALVE

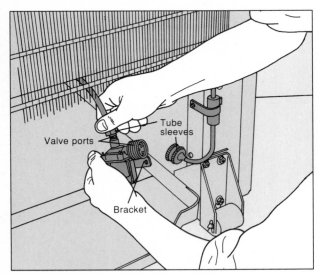

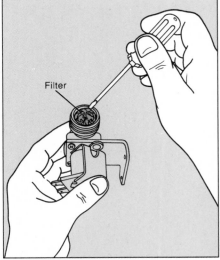

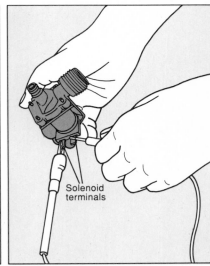

Inspecting and testing the water inlet valve. Unplug the refrigerator and turn off the water supply to the icemaker (usually at a valve under the sink). Pull the refrigerator out from the wall. To remove the water inlet valve, unscrew its bracket from the back of the refrigerator and disconnect the tube sleeves from the valve ports *(above, left)*. Keep a pan handy to catch dripping water. Disconnect the valve terminal plug and the ground wire. Use a screwdriver to pry out the filter from the inlet port *(above, center)* and rinse it in clear water. To test the water inlet valve solenoid, set a multitester at RX10 and touch a probe to each terminal *(above, right)*. The tester should show 200 to 500 ohms of resistance; if not, replace the water inlet valve. To install a new valve, reconnect the terminal plug and ground wire, screw the bracket on the refrigerator and reconnect the tube sleeves to the valve ports. Push the refrigerator back in position, plug it in and turn on the water. After installing a new water inlet valve, discard the first two or three batches of ice.

FREEZERS

Freezers are very similar to refrigerators in design and operation. Some even share the special features—automatic defrost, energy-saver switch—of top-of-the-line refrigerators. Upright freezers, especially, may have door hinges, gaskets, leveling feet, drainage systems, condenser fans and temperature controls more similar to those of a refrigerator than to the chest freezer pictured below. Consult the refrigerator Troubleshooting Guide *(page 15)* to diagnose problems with these components.

The evaporator coils and condenser coils of chest freezers are usually embedded in the cabinet, inaccessible to the home fixer. The door hinges and gasket differ from those of an upright model; so do the location of, and access to, the temperature control and compressor components.

The power indicator light, a safety feature common to most freezers, glows when there is sufficient current to the machine. The light goes off to alert you to an unplugged power cord, blown fuse or tripped circuit breaker, or general power outage. However, the light cannot tell you whether the freezer tempera-ture is low enough to keep food frozen. When in doubt, test the freezer temperature *(page 22)*; it should be about 0°F.

If the power goes out, food will keep (though not necessarily frozen) for 24 to 36 hours in a closed freezer. To store food longer, pack the freezer with dry ice (frozen carbon dioxide), protecting the food with a blanket or newspapers. If you must remove food briefly to work on the machine, place it in the bathtub, layered with newspaper and dry ice. Do not allow the dry ice to touch the food.

Defrost the freezer as needed, normally once or twice a year. Unplug the machine, take out the food and allow the ice to melt. To speed the job, use a hair dryer or place a pan of hot water in the freezer and close the door. Wash the freezer interior with a solution of baking soda and water. For better efficiency and longer life, do not locate a freezer near a heat source such as a basement furnace, and never put hot foods in the freezer. Always disconnect the power and discharge the capacitor, if any, before starting a repair.

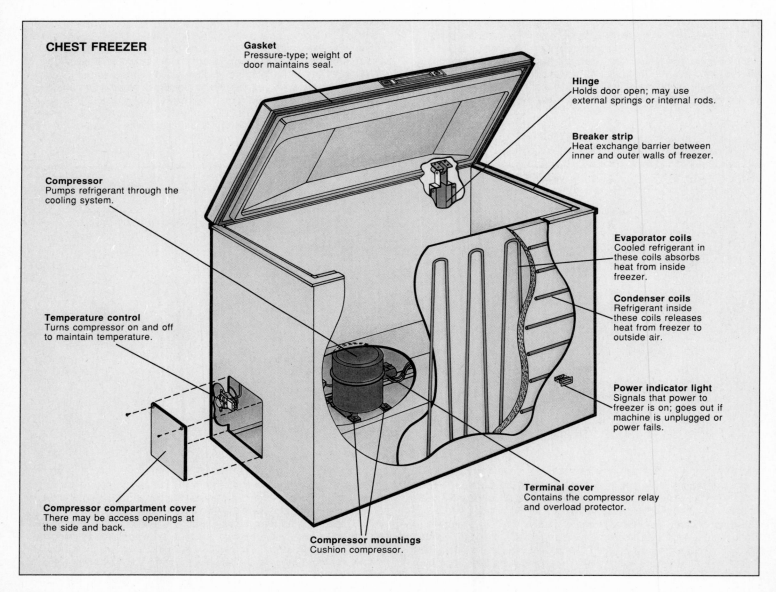

CHEST FREEZER

Gasket
Pressure-type; weight of door maintains seal.

Hinge
Holds door open; may use external springs or internal rods.

Breaker strip
Heat exchange barrier between inner and outer walls of freezer.

Compressor
Pumps refrigerant through the cooling system.

Evaporator coils
Cooled refrigerant in these coils absorbs heat from inside freezer.

Condenser coils
Refrigerant inside these coils releases heat from freezer to outside air.

Temperature control
Turns compressor on and off to maintain temperature.

Power indicator light
Signals that power to freezer is on; goes out if machine is unplugged or power fails.

Compressor compartment cover
There may be access openings at the side and back.

Terminal cover
Contains the compressor relay and overload protector.

Compressor mountings
Cushion compressor.

TROUBLESHOOTING GUIDE

SYMPTOM	POSSIBLE CAUSE	PROCEDURE
Freezer doesn't run; power indicator light off	No power to freezer	Check that freezer is plugged in; check for blown fuse or tripped circuit breaker *(p. 132)* □○
	Power cord loose or faulty	Inspect power cord *(p. 133)* ▣●▲
Freezer runs; power indicator light off	Light faulty	Test power indicator light *(p. 40)* ▣●
	Motor relay faulty	Test motor relay *(p.38)*. ▣●▲
Freezer starts and stops rapidly	Overload protector tripping repeatedly	Have electrician check voltage at outlet
	Compressor faulty	Test compressor *(p. 39)* ▣●▲
Freezer not cold enough; power indicator light on	Door opened too often	Open door less frequently
	Room too warm	Adjust room temperature
	Temperature control set too high	Set temperature control to lower setting
	Motor relay faulty	Test motor relay *(p. 38)* ▣●▲
	Overload protector faulty	Test overload protector *(p. 39)* ▣●
	Compressor faulty	Test compressor *(p. 39)*, ▣●▲ or call for service
	Temperature control faulty	Test temperature control *(p. 40)* ▣●
	Poor seal around door	Check breaker strip and gasket *(p. 41)* □○
	Refrigerant leaking or contaminated	Call for service
Freezer too cold	Temperature control set too low	Set temperature control to higher setting
	Temperature control faulty	Test temperature control *(p. 40)* ▣●
Moisture around freezer door or frame	Food not stored properly	Wrap food securely; see Use and Care manual
	Door opened too often	Open door less frequently
	Freezer in humid location	Move freezer or install dehumidifier
	Poor seal around door	Check breaker strips and gasket *(p. 41)* □○
Freezer runs constantly	Frost buildup	Defrost freezer
	Poor seal around door	Check breaker strips and gasket *(p. 41)* □○
Freezer smells bad	Insulation absorbing moisture	Remove breaker strips and allow insulation to dry; replace strips if damaged *(p. 41)* □○
Freezer noisy	Freezer not level	Adjust leveling feet, move freezer to level area of floor or put wooden shims under cabinet corners to level machine
	Compressor mountings loose or hardened	Replace compressor mountings *(p. 30)* ▣●
Door doesn't stay open	Hinge broken	Replace hinge *(p. 41)* ▣●

DEGREE OF DIFFICULTY: □ **Easy** ▣ **Moderate** ■ **Complex**
ESTIMATED TIME: ○ **Less than 1 hour** ◐ **1 to 3 hours** ● **Over 3 hours**　　　　▲ **Multitester required**

ACCESS TO THE COMPRESSOR COMPARTMENT

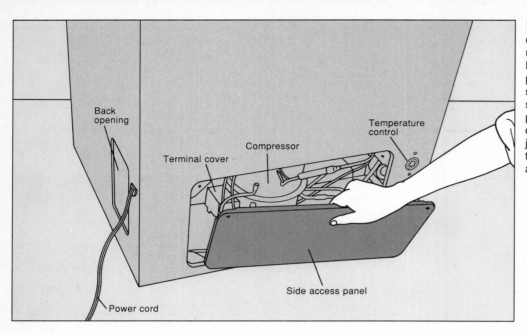

Back opening

Terminal cover

Compressor

Temperature control

Power cord

Side access panel

Reaching the interior components. Most chest freezers have an opening in the back, usually without a cover. Some models also have a side opening, covered by an access panel *(left)*. Unplug the freezer. Remove the screws from the panel and lift it off, as shown. You now have access to the compressor and terminal cover, the temperature control and the power cord. To perform tricky jobs such as servicing the temperature control, use both openings to give your hands access from two angles.

TESTING THE COMPRESSOR RELAY

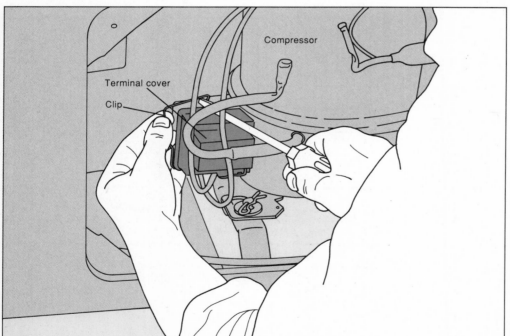

Compressor

Terminal cover

Clip

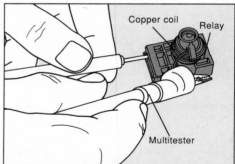

Copper coil

Relay

Multitester

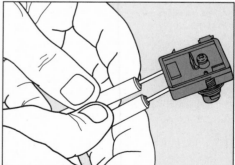

Testing the relay. To reach the compressor relay, as well as the overload protector and compressor, unplug the freezer and remove the access panel, if any. Use a screwdriver to pry off the clip holding the terminal cover to the compressor *(above, left)*, and take off the cover. The relay plugs onto the compressor pins; pull it straight off without jiggling it. Label and disconnect the wires.

If the relay has a copper coil, hold it with the word TOP facing up. Set a multitester at RX1; clip one probe to terminal L (on the side of the relay) and insert the other probe into terminal M *(right, top)*.

The tester should show continuity. Remove the probe from terminal M and insert it into terminal S; the tester should show resistance. Turn the relay upside down; you should hear a click and the tester should now show continuity. Move the probe from terminal L to terminal M *(right, bottom)*; the tester should show continuity.

If the relay has no copper coil, insert a multitester probe into each terminal; the tester should show 5 to 10 ohms of resistance. If either type of relay fails any of these tests, replace it.

TESTING THE OVERLOAD PROTECTOR

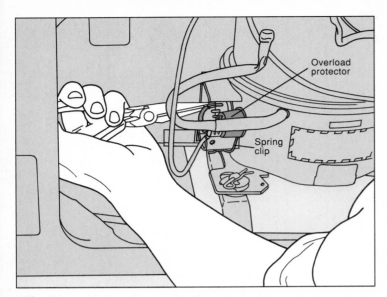

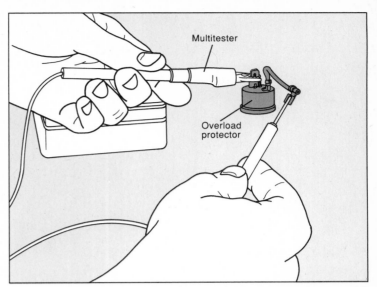

1 **Disconnecting the overload protector.** To reach the protector, unplug the freezer and remove the relay *(page 38)*. Label and disconnect the wires to the overload protector *(above)*, then disengage its spring clip and lift out the protector. Allow it to cool to room temperature before testing.

2 **Testing the overload protector.** With a multitester set at RX1 (or a continuity tester), touch one probe to each terminal of the protector *(above)*. The tester should indicate continuity. If the protector tests OK, next check the compressor *(below)*. To replace a faulty overload protector, first reinstall the relay. Insert the new protector into the spring clip, reconnect the wires and replace the terminal cover and access panel.

TESTING THE COMPRESSOR

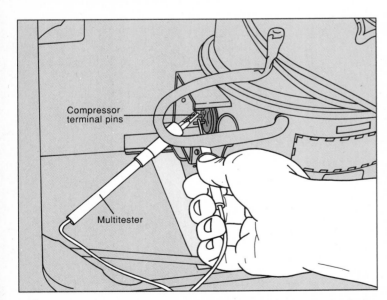

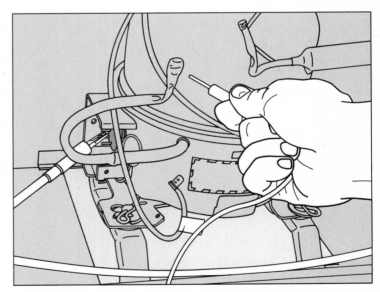

1 **Testing the compressor.** To access the compressor terminal pins, unplug the freezer and remove the terminal cover, relay and overload protector. If the compressor is hot, allow it to cool to room temperature. With a multitester set at RX1, test each compressor terminal pin against the other two pins. Each pair should show 1 to 10 ohms of resistance; if not, call for service.

2 **Testing for a ground.** Set a multitester at RX1000. Touch one probe to the bare metal of the compressor housing (if necessary, scrape a little paint off the housing for a good contact). Touch the other probe to each compressor pin in turn. If the tester needle moves at all, the compressor is grounded; call for service. If the compressor tests OK and still doesn't work, it may have a mechanical problem; call for service. To reassemble the freezer, reinstall the overload protector, relay and terminal cover, and replace the access panel.

TESTING THE TEMPERATURE CONTROL

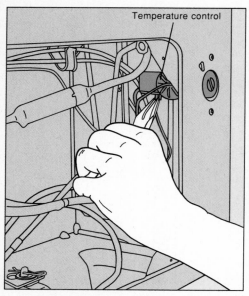

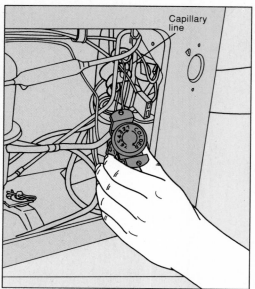

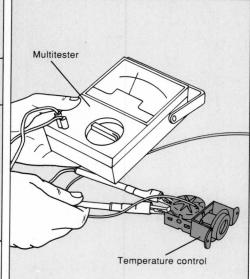

Servicing the temperature control. Unplug the freezer and remove the access panel. Disconnect the wires from the temperature control terminals *(above, left)* and label their positions for reassembly. Unscrew the control from the freezer cabinet and carefully pull it out *(above, center)*, using your other hand to ease the capillary line out of its channel without kinking it. Touch a tester probe to each control terminal *(above, right)*. With the control at its coldest setting, the tester should show continuity. With the control at its highest setting (or OFF), the tester should show resistance. To replace a faulty temperature control, carefully thread the capillary line into its channel, screw the new control to the freezer and replace the access panel.

TESTING THE POWER INDICATOR LIGHT

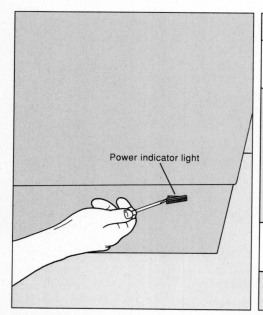

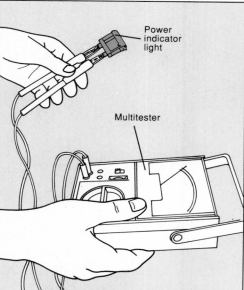

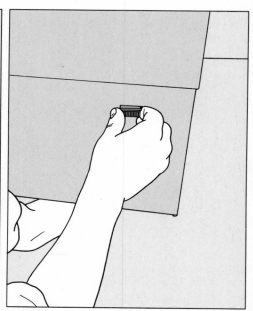

Removing and testing the light. Unplug the freezer. Using a small screwdriver, gently pry out the power indicator light from the front of the freezer cabinet *(above, left)*. Label and disconnect the wires, and tape them to the cabinet so that they don't slip back into the machine. Touch one probe of a multitester set at RX1 (or a continuity tester) to each terminal *(above, center)*; the tester should indicate continuity. To replace the light, connect the wires to a new light and snap it into the cabinet. If the wires fell back through the hole, prop up the freezer on a block of wood and reach up under the bottom to retrieve them *(above, right)*.

REMOVING THE DOOR HINGES

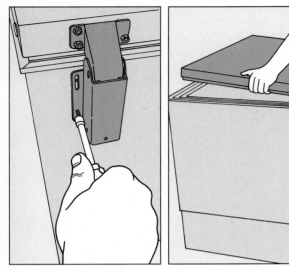

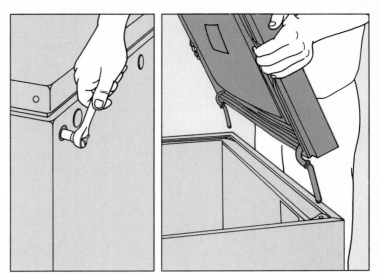

Removing a spring-type hinge. Unplug the freezer and remove the screws securing each hinge to the freezer cabinet *(above, left)*. Hold the hinge so that it doesn't snap up. Do not remove the hinge screws from the freezer door. Lift the door and hinges off the freezer cabinet *(above, right)*. To replace the door, align the hinges against the cabinet, and replace and tighten the screws.

Removing a rod-type hinge. Unplug the freezer and use a socket wrench to unscrew the bolts holding each hinge to the freezer cabinet *(above, left)*. Lift the door and hinges off the cabinet *(above, right)*. To replace the door, insert the hinge rods in the holes, align the door with the freezer cabinet and tighten the hinge bolts.

CHANGING THE BREAKER STRIPS AND GASKET

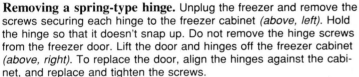

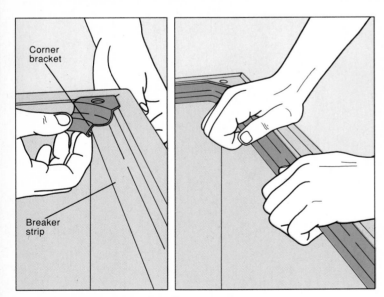

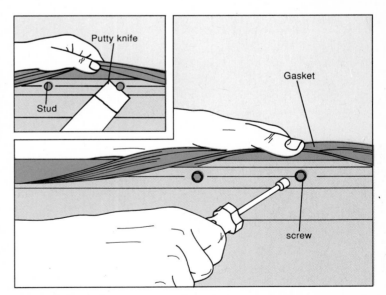

Replacing breaker strips. Unplug the freezer, open the door and cover the contents with newspapers or a blanket. Snap out the corner bracket at each end *(above, left)*; if stiff, use a hair dryer set on LOW to warm them slightly. Snap out the breaker strip with your hands *(above, right)* or pry it off with a putty knife padded with masking tape. If the insulation below is dry, snap in a new breaker strip and replace the corner brackets. If damp, leave it uncovered to dry, and remove the freezer contents to cold storage.

Replacing the gasket. Unplug the freezer. Remove the door, spread a blanket over the freezer and lay the door upside down on top of it. Soak the new gasket in warm water to soften it. If the gasket is held in place by plastic studs *(inset)*, use a putty knife padded with masking tape to pry them out, three or four at a time, inserting the new gasket as you go.

If the gasket is screwed in place *(above)*, remove several screws at a time, pull out the old gasket and insert the new one. Replace the screws loosely, then tighten them all. Reinstall the door.

ELECTRIC RANGES

Although the heart of an electric range seems a tangle of wires and switches, most repairs are based on simple deductive reasoning. A electric range operates on a 240/120-volt circuit—240 volts for the heating elements and 120 volts for the accessories (clock, lights and appliance receptacle). The range draws power though two separate fuses or breakers; if you suspect an electrical problem, check both fuses or breakers first.

The heating elements are controlled by electrical switches. A thermostat senses and regulates oven temperature. Self-cleaning ovens use extremely high temperatures—about 900°F—to burn food residue off the oven walls. A special door-lock mechanism prevents the oven door from being opened until the cleaning cycle is completed.

Electric ranges are made in a variety of styles—freestanding, slide-in, double-oven, cooktop or wall oven—but all operate in much the same way and use similar components. Once you understand the basic repair procedures, you can adapt them to your own range. Most of the repairs on the following pages are shown with a freestanding range.

The Troubleshooting Guide at right lists the most common malfunctions in order from most to least likely. Before deciding that your range needs repair, check that the problem is not due to incorrect use. Follow the use and care recommendations in your owner's manual. Keeping your range clean is the most effective way to avoid breakdowns, but be careful not to get cleaning liquids inside the range where they can cause short

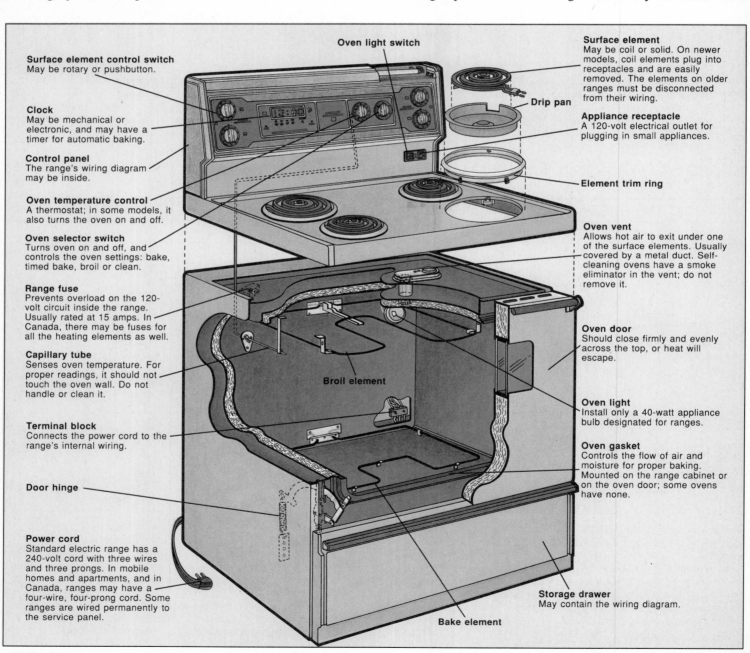

Surface element control switch
May be rotary or pushbutton.

Clock
May be mechanical or electronic, and may have a timer for automatic baking.

Control panel
The range's wiring diagram may be inside.

Oven temperature control
A thermostat; in some models, it also turns the oven on and off.

Oven selector switch
Turns oven on and off, and controls the oven settings: bake, timed bake, broil or clean.

Range fuse
Prevents overload on the 120-volt circuit inside the range. Usually rated at 15 amps. In Canada, there may be fuses for all the heating elements as well.

Capillary tube
Senses oven temperature. For proper readings, it should not touch the oven wall. Do not handle or clean it.

Terminal block
Connects the power cord to the range's internal wiring.

Door hinge

Power cord
Standard electric range has a 240-volt cord with three wires and three prongs. In mobile homes and apartments, and in Canada, ranges may have a four-wire, four-prong cord. Some ranges are wired permanently to the service panel.

Oven light switch

Surface element
May be coil or solid. On newer models, coil elements plug into receptacles and are easily removed. The elements on older ranges must be disconnected from their wiring.

Drip pan

Appliance receptacle
A 120-volt electrical outlet for plugging in small appliances.

Element trim ring

Oven vent
Allows hot air to exit under one of the surface elements. Usually covered by a metal duct. Self-cleaning ovens have a smoke eliminator in the vent; do not remove it.

Oven door
Should close firmly and evenly across the top, or heat will escape.

Oven light
Install only a 40-watt appliance bulb designated for ranges.

Oven gasket
Controls the flow of air and moisture for proper baking. Mounted on the range cabinet or on the oven door; some ovens have none.

Broil element

Bake element

Storage drawer
May contain the wiring diagram.

circuits. Don't use foil to line the drip pans under the burners or oven element—it can short the electrical connections. Using burners without drip pans can also harm the wiring. Never wash the gasket of a self-cleaning oven.

Most range repairs are electrical in nature, but they are not complex. Many malfunctions are caused by loose connections or burned wires; always check for these first. Clues to a loose connection are a metallic odor or a soft hissing or buzzing. A sharp odor of burning plastic indicates overheating in a switch or terminal block. When replacing wires, use the same gauge insulated wire used by the manufacturer. For some repairs, you must refer to your range's circuit diagram, located on the back panel, in the storage drawer or inside the control panel.

Before starting repairs, unplug the range or turn off the power at the service panel. Check that you've disconnected the right fuses or breakers by turning on the heating elements—they should not warm up. While unplugging the range, don't touch the back panel; a loose wire inside could shock you. Before reconnecting the power, make sure that no uninsulated wires or terminals touch the cabinet, and that wiring is away from sharp edges and moving parts.

Many ranges, electric or gas, are topped by a venting range hood. Most range hoods have a fan or a "squirrel-cage" blower wheel that pulls smoke and grease through an aluminum mesh filter and out an exhaust duct. Common range hood problems are listed at the end of the Troubleshooting Guide.

TROUBLESHOOTING GUIDE
continued ▶

SYMPTOM	POSSIBLE CAUSE	PROCEDURE
Nothing works	No power to range	Check fuses or circuit breakers (p. 132) □○
	Power cord faulty	Test power cord (p. 133) ■●
	Terminal block burned	Check terminal block (p. 133) ■●
All elements do not heat, or heat only partially	Partial power to range	Check fuses or circuit breakers (p. 132) □○
	Poor connection at terminal block	Check terminal block (p. 133) ■●
	Power cord faulty	Test power cord (p. 133) ■●
Surface element doesn't heat	Loose connection at element terminals	Reposition element (p. 46) □○
	Element shorted	Test element (plug-in elements, p. 46 □○; wired and solid-disc elements, p. 47 ■●▲)
	Receptacle faulty	Check receptacle (p. 46) ■●▲
	Burner switch faulty	Test switch (p. 48) ■●▲
Surface element provides only high heat	Burner switch faulty	Replace switch (p. 48) ■●▲
Oven doesn't heat	Clock timer incorrectly set	Reset clock timer □○
	Bake or broil element faulty	Test element (p. 49) ■●▲
	Temperature control faulty	Test temperature control (p. 51) ■●▲
	Oven selector switch faulty	Test switch (p. 52) ■●▲
Oven doesn't hold set temperature	Capillary tube broken or touching oven wall	Check capillary tube (p. 50) ■●
	Oven door not aligned	Adjust door (p. 55) □○
	Oven gasket broken	Replace oven gasket (cabinet-mounted ovens, p. 56 □○; self-cleaning ovens, p. 57 ■●)
	Temperature control out of adjustment	Test oven temperature, recalibrate control (p. 50) ■●
	Temperature control faulty	Test temperature control (p. 51) ■●▲
Oven produces condensation	Oven vent clogged	Clean vent, wash duct (p. 50) □○
Self-cleaning oven doesn't clean	Oven door unlocked	Reclose and lock door
	Bake or broil element faulty	Test element (p. 49) ■●▲
	Clock timer faulty	Check clock (p. 52) ■●
	Temperature control faulty	Test temperature control (p. 51) ■●▲
	Oven selector switch faulty	Test switch (p. 52) ■●▲
	Door lock broken	Call for service
	Smoke eliminator faulty	Call for service
Clock or timer doesn't work	Range fuse blown	Check range fuse (p. 54) □○
	Clock faulty	Check clock (p. 52) ■●
Oven light out	Bulb loose or burned	Replace bulb (p. 53) □○
	Light switch broken	Test switch (p. 53) ■●▲
Appliance receptacle doesn't work	Range fuse blown	Check range fuse (p. 54) □○
	Receptacle faulty	Check receptacle (p. 54) □○

DEGREE OF DIFFICULTY:	□ Easy ■ Moderate ■ Complex	
ESTIMATED TIME:	○ Less than 1 hour ● 1 to 3 hours ● Over 3 hours	▲ Multitester required

TROUBLESHOOTING GUIDE (continued)

SYMPTOM	POSSIBLE CAUSE	PROCEDURE
Oven door doesn't close properly	Oven door not aligned	Adjust door *(p. 55)* □○
	Spring loose or broken	Adjust or replace springs *(p. 55)* □○
	Hinge broken	Call for service

RANGE HOODS

SYMPTOM	POSSIBLE CAUSE	PROCEDURE
Fan or blower doesn't run	No power to hood	Check fuse or circuit breaker *(p. 132)* □○
Smoke or smell lingers in the kitchen	Filter dirty	Clean or replace filter *(fan-type range hood, p. 58* □○*; squirrel-cage type, p. 59* □○*)*
	Ductwork too small or blocked	Call for service
Grease on range top	Dirty filter drips grease	Clean or replace filter *(pp. 58, 59)* □○
Fan blade rotation slow or uneven (fan-type range hood)	Dirt and grease buildup on fan blades	Clean fan *(p. 58)* □○
	Motor shaft binds	Lubricate shaft *(p. 58)* □○
Fan or blower not exhausting fumes	Filter dirty	Clean or replace filter *(pp. 58, 59)* □○
	Hood not positioned correctly over range	Check to see that hood covers entire cooking area.
	Blower assembly mounted incorrectly (squirrel-cage type)	Remount blower assembly *(p. 59)* ◪◐
	Ductwork blocked by dirt and grease	Call for service
Hood vibrates	Blower wheels out of alignment (squirrel-cage type)	Replace wheels *(p. 59)* ◪◐

DEGREE OF DIFFICULTY: □ Easy ◪ Moderate ■ Complex
ESTIMATED TIME: ○ Less than 1 hour ◐ 1 to 3 hours ● Over 3 hours   ▲ Multitester required

ACCESS TO THE RANGE

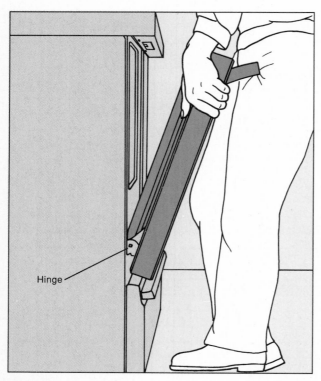

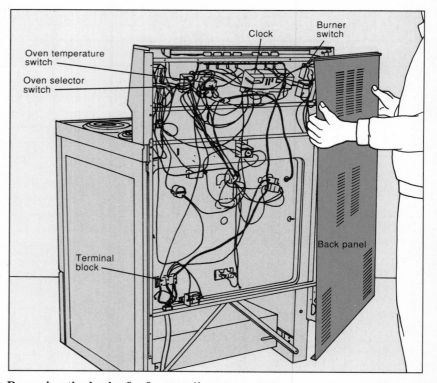

Removing the oven door. Many oven doors slide off their hinges, facilitating work inside the oven or on the door itself. (The door hinges of a self-cleaning range must be unscrewed). Open the door to its first stop, the broil position. Grip each side as shown, maintaining the door's angle, and pull the door straight off it hinges. Close the hinge arms against the oven for safety.

Removing the back of a freestanding range. The wiring and controls of a free-standing range are reached from the back. Unplug the range or turn off power at the service panel. Rock the range away from the wall. Supporting the back panel with a free hand or knee, remove the screws from around the panel's edges. Some ranges have a single panel *(above)*; others have a lower panel covering the terminal block where the power cord is attached, and one or more upper panels covering the wiring and controls.

ACCESS TO THE RANGE (continued)

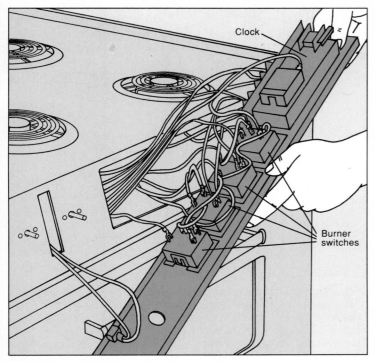

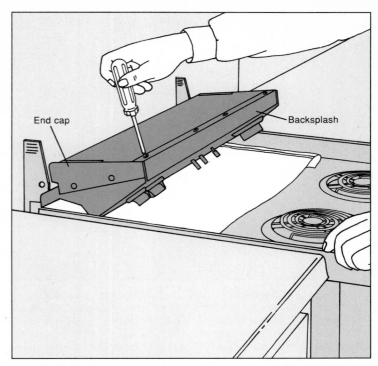

Reaching front-mounted controls. This type of control panel is attached with screws at each end. After disconnecting power to the range, remove them with a screwdriver. The panel may be held by a spring clip as well; pull up the panel to release it from the clip. Tilt the panel forward to expose the controls and wiring *(above)*; rest it on the top of the range or the edge of the door while working.

Removing the backsplash of a built-in range. Although similar in style to a freestanding range, the controls of a built-in range often can be serviced from the front, without moving the range from the wall. Disconnect power to the range and remove the screws from each end cap of the backsplash. Spread a towel on the cooktop to protect it. Pull the backsplash forward and rest it on the towel. To reach the range switches and controls, unscrew the rear panel of the backsplash, as shown, and set it aside.

Removing the control panel from an upper oven or a wall oven. Disconnect power to the appliance and remove the screws at each end of the control panel. (You may have to open the door of a recessed wall oven to remove the screws from inside.) Pull out the oven rack, ease the panel forward, and rest it on the rack. If the control panel is hinged at the bottom, simply open it toward you. If your upper oven has a range hood built above it, you may have to move it out of the way to free the panel, by releasing a latch pin under the hood.

REPAIRING PLUG-IN BURNER ELEMENTS

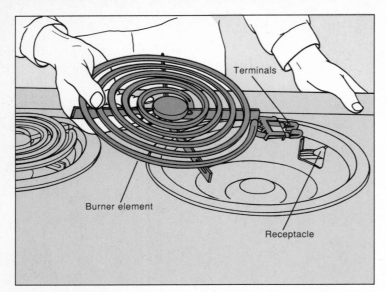

Terminals

Burner element

Receptacle

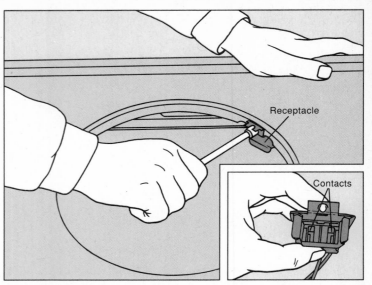

Receptacle

Contacts

1 **Checking a plug-in burner element.** Most modern ranges have sheathed coil elements with terminals that plug into an electrical receptacle within the burner opening. If a plug-in element doesn't heat, disconnect power to the range, grasp the element and reseat its terminals securely in the receptacle. If the problem persists, lift the element up about an inch and pull it out, as shown. Inspect the element for damage; if the terminals are corroded, buff them with fine steel wool and reinstall the element. If it is burned or pitted, replace it. If the element shows no visible damage, test it by plugging it into the receptacle of a working element; if it still does not heat, replace it. If the element does heat, check its receptacle.

2 **Examining an element receptacle.** Lift out the drip pan and its chrome ring. Unscrew the receptacle from the range, as shown. Pull out the receptacle, taking care not to strain the wiring, and examine the metal contacts inside *(inset)*. If the contacts appear bent, burned or oxidized, replace the receptacle. Examine the terminals at the back of the receptacle where the wires are connected (you may first have to snap off two clips and remove a card-like insulator). If the wire terminals are burned, cut them off and splice on new terminals *(page 136)*. On some ranges, you must lift and prop the cooktop to work on the receptacles. If screws secure the front edge of the cooktop, remove these before raising the top.

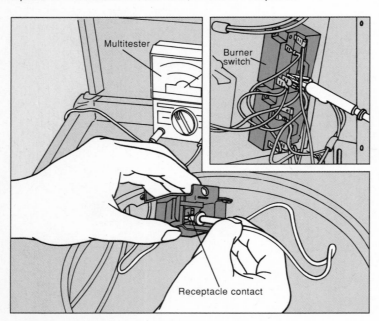

Multitester

Burner switch

Receptacle contact

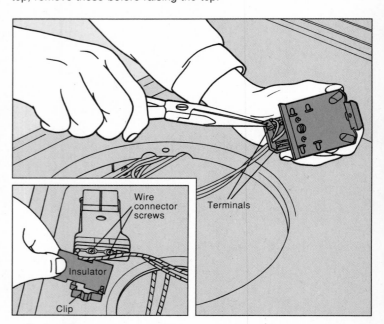

Wire connector screws

Terminals

Insulator

Clip

3 **Testing the receptacle.** Since damage to a receptacle is not always visible, test it for continuity. With the power disconnected, gain access to the range switches and controls *(page 44)*. Trace the wires from the receptacle to the corresponding terminals on the burner switch (usually marked H1 and H2). Clip one probe of a multitester to a terminal *(inset)*, and touch the other to each of the receptacle contacts in turn, as shown. Only one contact should show continuity. Repeat with a probe on the second switch terminal. The other contact should show continuity.

4 **Replacing the receptacle.** Use long-nose pliers to pull the wire connectors from the receptacle terminals, as shown. To install a new receptacle, reattach a connector wire to each terminal and screw the receptacle firmly to the range. On some models, you must cut the wires leading to the receptacle and splice a new one in place *(page 136)*. If the receptacle's wiring connections are covered by a card-like insulator, snap off the clip with a screwdriver to remove the insulator, then unscrew the wires *(inset)*. Screw the wires to the new receptacle and clip on the insulator.

REPAIRING WIRED BURNER ELEMENTS

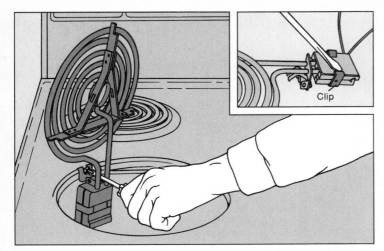

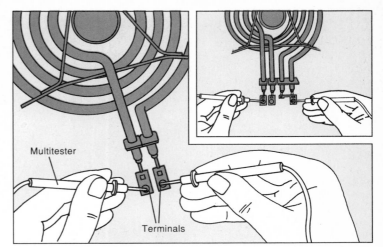

1 **Disconnecting a wired element.** The burners on some ranges, rather than plugging into a receptacle, are connected directly to the burner switch wires; this connection is protected by a glass or ceramic block. Lift the element and inspect its coils for burns or holes, and replace if damaged. To remove the element, turn off power to the range, remove the drip pan and unscrew the element and block from the range *(above)*. Use a screwdriver to pry off the clips joining the two halves of the block *(inset)*, and pull them apart to expose the screws that connect the wires. Tighten loose connections or repair burned wiring *(page 136)*. To test the element, unscrew the wires without bending the terminals. If there are more than two wires, label their positions with masking tape.

2 **Testing a wired element.** With a multitester set at RX1, touch one probe to each element terminal *(above)*; the meter should show only partial resistance. Next, test for a ground with one probe on a terminal and the other on the coil sheathing. The multitester needle should not move. If the element has several terminals, half of them lead to a common terminal. Clip one probe to the common terminal and touch the other to each terminal in turn *(inset)*; the multitester should show partial resistance. Test for a ground by touching one probe to the coil sheathing and the other to each terminal in turn; there should be no continuity. To install a new element, screw the wires to the proper terminals, clip the insulating block in place and screw the element securely to the range.

REPAIRING SOLID-DISC BURNER ELEMENTS

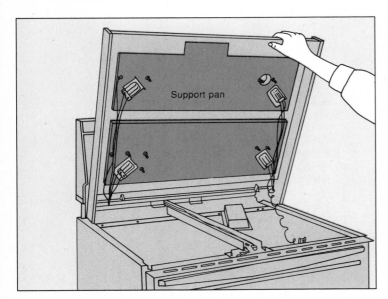

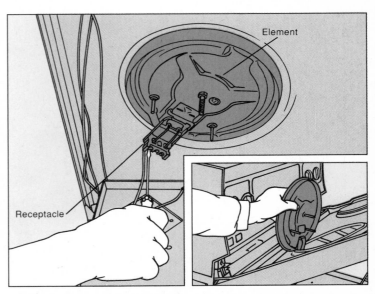

1 **Accessing a solid-disc element.** Disconnect power to the range and lift and prop the cooktop. If there are screws under the front edge of the range top, behind the oven door, remove them before raising the top. Remove the nuts or screws securing the support pan or bracket under the element; you may also have to disconnect a green or copper ground wire. Set the bracket aside.

2 **Testing a solid-disc element.** Unscrew the wires from the element's terminals, as shown, and pull the element out through the top of the range *(inset)*. Set a multitester at RX1 and touch a probe to each terminal. The needle should sweep partially upscale, showing some resistance. Test for a ground by placing one probe on a terminal and the other on the element's metal sheath; the multitester needle should not move. If the element fails either test, replace it. Insert the new element through the top, screw the wires firmly to the terminals, and replace the support pan or bracket.

TESTING AND REPLACING BURNER ELEMENT SWITCHES

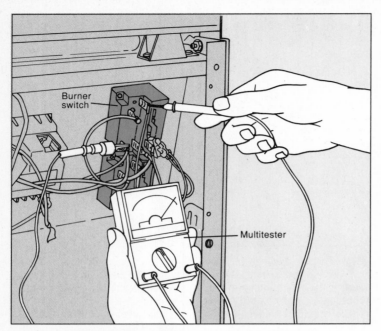

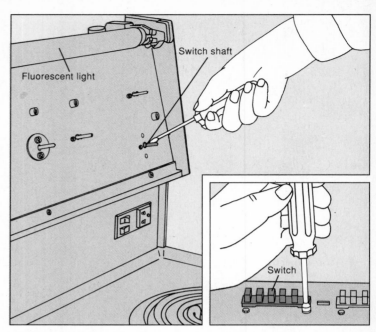

1 **Testing a burner switch.** Disconnect the power and gain access to the range controls and switches *(page 44)*. Replace any switch with an obvious mechanical problem, such as a jammed button *(step 2)*. Check the switch wires for loose connections. If the switch terminals are damaged, replace the switch and any burned wire terminals *(page 136)*. If there is no visible damage, test the switch for continuity. The power-supply wires are attached to terminals marked L1 and L2 (and a terminal marked N on some switches). Wires leading to the burner element are marked H1 and H2 (or just numbered). Turn on a working switch and test each power-supply terminal, in turn, against each burner-element terminal in the switch position that does not heat; disconnect one wire in each pair being tested. Then test the suspect switch; if the results don't match, replace it.

2 **Removing the switch.** To remove a rotary switch, pull off the control knob and remove the two screws holding the switch to the control panel, as shown. Pull the switch out through the back. If a glass panel covers the screws, first pull off all the control knobs and check for clips or trim pieces at the top or sides of the panel (you may have to raise the fluorescent light cover). Unclip the panel or unscrew the trim pieces, and lift out the panel. Pushbutton switches may be trimmed by a removable panel that covers the mounting screws; unfasten the clips holding the panel and pry it off. Remove the switch screws *(inset)* and lift out the switch through the back.

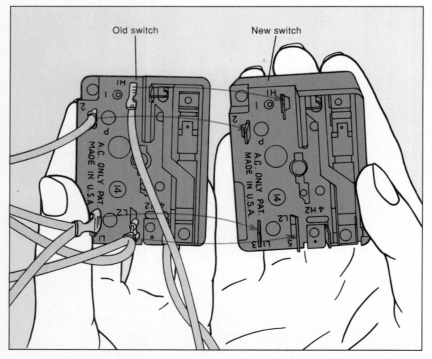

3 **Replacing the switch.** Leave the wires connected to the old switch while you buy a replacement part with exactly the same part number (if you leave the control panel disassembled, leave a note taped to it warning other members of the household not to reconnect the power).
　Replace the switch in one of two ways: Transfer the wires from the old switch to the same terminals on the new switch one by one. Or, after labeling the position of each wire with masking tape, disconnect all the wires from the old switch, and connect them to the new switch. Screw the new switch to the control panel and replace the control panel cover, trim and control knobs.

REPLACING OVEN ELEMENTS

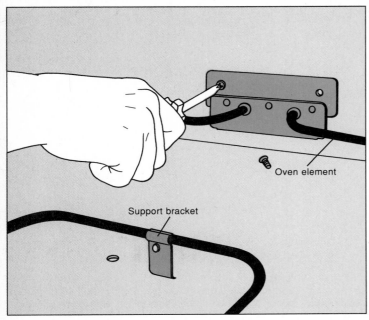

Oven element

Support bracket

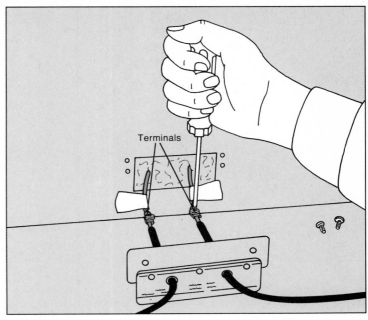

Terminals

1 **Dismounting the oven element.** Bake and broil elements are tested and replaced in the same way. Disconnect power to the range. For easy access to the oven, remove the door *(page 44)*. Remove the screws or nuts that fasten the element to the back of the oven *(above)*. The element may also have a front support bracket; unscrew it if necessary. Gently pull the element forward a few inches to expose its wiring. If the capillary tube is in the way of the broil element, unclip the tube from its support without bending it *(page 51)*. **Caution:** In self-cleaning ovens, the tube contains a caustic fluid—wear rubber gloves and use caution when handling it.

2 **Disconnecting the element.** Label the wire positions with masking tape and unscrew them from the element terminals, as shown. Avoid bending the terminals. Do not allow the wires to fall back through the opening—otherwise, the range must be pulled out from the wall and opened to retrieve them. Check the wire connectors for burns; if damaged, cut them off and replace them *(page 136)*. Remove the element from the oven. Some older models have plug-in elements that can be pulled out rather than unscrewed.

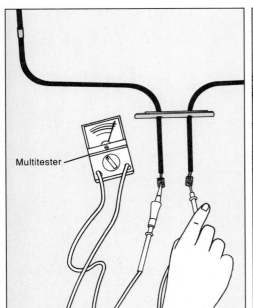

Multitester

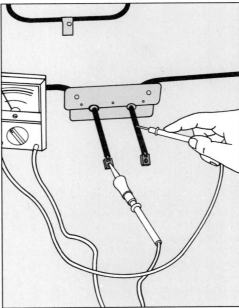

3 **Testing and replacing the oven element.** With a multitester set at RX1, touch a probe to each of the element terminals *(far left)*; there should be only partial resistance. If not, replace the element with a new one of the same wattage. Next, test for a ground with one probe on a terminal and the other on the metal sheath of the element *(near left)*; the multitester needle should not move. If the element fails either of these tests, replace it. To install a new element, reconnect the wires to the element terminals and screw the rear support bracket firmly in place for proper grounding. Reattach the support bracket if necessary and make sure the capillary tube is properly seated in its clips.

CHECKING AND ADJUSTING OVEN TEMPERATURE

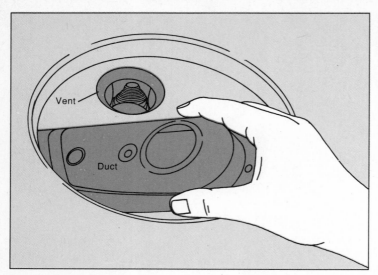

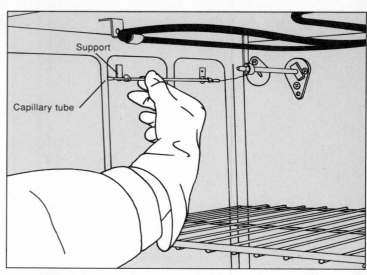

Checking and cleaning the oven vent. The vent helps control the circulation of air in the oven—and therefore the oven temperature—by conducting hot air through a duct under a burner, usually the right rear element. Disconnect power to the range. Pull out the burner, or raise it out of the way, and remove the drip pan. Lift out the vent duct as shown, to expose the vent—you may need to unscrew the duct first. Clean the vent and wash the duct in hot, soapy water. When replacing the duct, be sure its opening will line up with the hole in the drip pan before screwing it in position. Replace the drip pan and burner.

Checking the capillary tube. Clipped to the inside of the oven, the capillary tube senses oven temperature and is adjusted by the oven temperature control. If the tube touches the oven wall, reposition it in its support clips. If it is broken, both the tube and the temperature control switch must be replaced *(page 51)*. **Caution:** In self-cleaning ovens, the tube contains caustic chemicals. Wear rubber gloves and avoid bending the tube when handling it. If you should get the contents on your skin, rub it off with a dry towel before washing with mild soap and water.

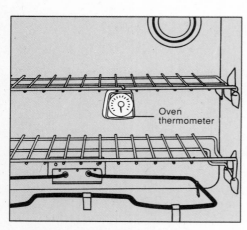

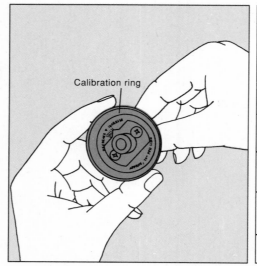

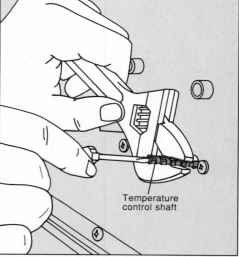

Testing oven temperature. Place an oven thermometer in the center of the oven and set the temperature control switch at 350°F . Wait 20 minutes, then check the thermometer. Take three more readings, one every 10 minutes. Add the readings and divide by four; the average should be 350°F. If your result is off by 25°F or less, the control is normal; by 25°F to 50°F, recalibrate it *(next step)*. If off by more than 50°F, replace the control *(page 51)*.

Calibrating the temperature control. Pull the knob off the oven temperature control. If the back of the knob has a ring with marks indicating "Raise" and "Lower", turn the knob to move the ring, *(above, left)*; you may first have to loosen two screws on the ring. If the calibration device is mounted on the range under the knob, turn off the power, then loosen the screws to adjust the shaft. Give the shaft one-eighth of a turn to the right to lower the temperature, or to the left to raise it. Check the temperature again after calibration. If calibration doesn't work, replace the temperature control *(page 51)*.

TESTING AND REPLACING THE OVEN TEMPERATURE CONTROL

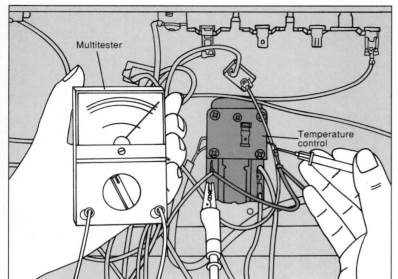

1 Testing the temperature control. Turn off power to the range and open the control panel *(page 44)*. If any of the temperature control terminals are discolored or burned, replace the temperature control. Next, test the control for continuity. If it has more than two terminals, refer to the wiring diagram—on the rear panel, or inside the storage drawer or control panel—for the correct pairs of terminals to test. Disconnect one wire of the pair, clip a tester probe to each terminal, and turn the switch to 300°F. If any of the circuits do not show continuity, replace the temperature control.

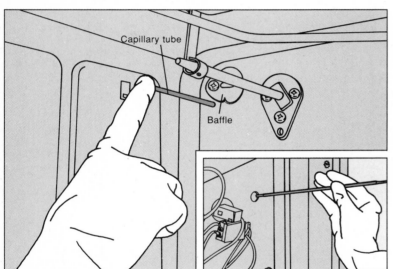

2 Removing the capillary tube. This tube leads from the temperature control into the oven. Gently unclip the tube from the supports in the oven and push it through the hole in the rear wall, as shown; you may have to loosen a screw that secures a baffle over the hole and slide the baffle aside. From the back of the range, pull the tube completely out of the oven *(inset)*.
Caution: In self-cleaning ovens, the tube is filled with caustic chemicals. Wear goggles and rubber gloves, and avoid bending the tube. If you should get the contents on your skin, rub it off with a dry towel before washing with mild soap and water.

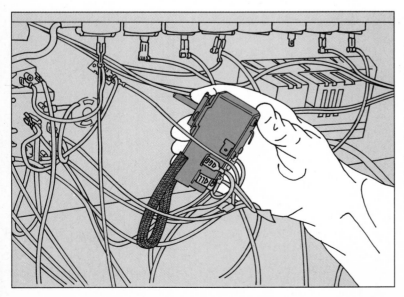

3 Removing and replacing the temperature control. Unscrew the two temperature control screws in front and remove the control from the back of the range, as shown. Label the positions of the wires and disconnect them from the control. Replace or splice any burned wire connectors *(page 136)*. To install a new temperature control, connect the wires to the terminals and screw the new switch to the control panel. Push the capillary tube gently into the oven through the the back, taking care not to bend or kink it, and clip it into its supports. Replace the rear panel, the front panel and the control knobs. Before replacing the knob, check that the calibration ring is centered. If not, reset it *(page 50)*.

TESTING AND REPLACING THE OVEN SELECTOR SWITCH

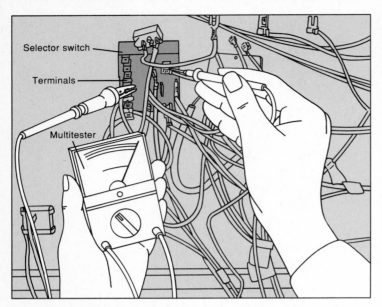

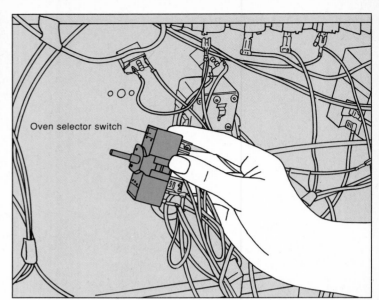

1 **Testing the oven selector switch.** Since it controls the bake, broil, timed bake and clean functions, a broken selector switch can cause any of these cycles not to work. Disconnect power to the range and open the control panel *(page 44)*. Replace the switch if any terminals are burned. Test the switch for continuity; the correct pairs of terminals to test for each switch setting are indicated on the wiring diagram. Disconnect one wire from each pair of terminals being tested, and check for continuity at each position of the switch, as shown. Replace the switch if it fails the test.

2 **Replacing the oven selector switch.** Remove the screws from the front of the control panel and pull the switch out from the back, as shown. Label the wires and disconnect them. Replace or splice any burned or corroded wire connectors *(page 136)*. To install a new switch, connect the wires to the terminals and screw the switch securely to the control panel to ensure proper grounding. Reassemble the control panel.

CHECKING AND REPLACING THE CLOCK

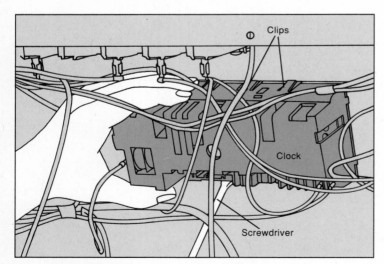

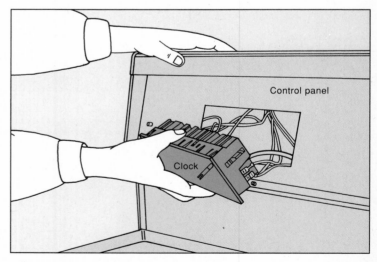

1 **Dismounting the clock.** Disconnect power to the range and open the control panel *(page 44)*. Check the wires to the clock terminals for loose connections. In many models, the clock is held to the back of the control panel by spring clips. Use the flat edge of a screwdriver to push in the clips, as shown. Screws or nuts may hold the clock to the back of the range; on some models, you must remove the temperature control switch and selector switch mounting screws, which fasten the clock bracket to the control panel. On others, you must unscrew the control panel to reach the clock mounting screws or nuts.

2 **Replacing the clock.** Remove the clock by pulling it out through the front of the range, as shown. (On some models, the clock is taken out through the back.) Label and disconnect the wires, or transfer the wires to the new clock one by one. Install a new clock by pushing it through the front of the control panel and snapping or screwing it into place. Screws must be attached securely to ensure proper grounding. Reassemble the control panel.

SERVICING THE OVEN LIGHT, SWITCH AND SOCKET

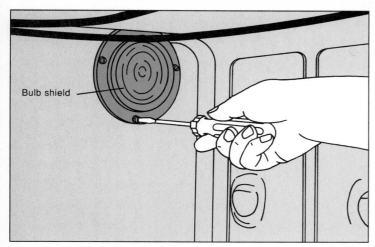

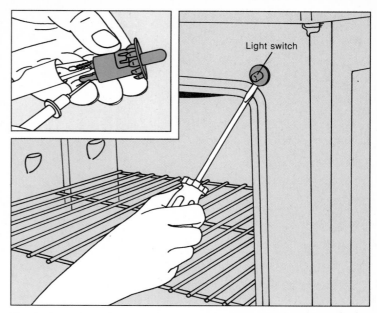

Replacing the oven light bulb. In ovens with a wire protector or a glass shield covering the bulb, first disconnect power to the range. Pull down the wire protector, or unscrew the glass shield, as shown. Unscrew the bulb, using a dry cloth to protect your hand. If the bulb breaks in the socket, be sure the power is disconnected before removing the remnants with long-nose pliers. Screw in a regular 40-watt bulb; if it works, the problem is simply a burned-out bulb. Replace it with an appliance bulb of the same size and wattage. If not, check the switch or the socket.

Testing and replacing a door-operated light switch. Disconnect power to the range and open the oven door. Pry out the switch, pushing in the spring clips with a screwdriver *(above)*. To test the switch, disconnect the wires and clip a continuity tester to each terminal *(inset)*. With the switch plunger in, there should be no continuity; when released, the switch should show continuity. If it fails the test, replace the switch. Attach the wires to the new switch and snap it into place.

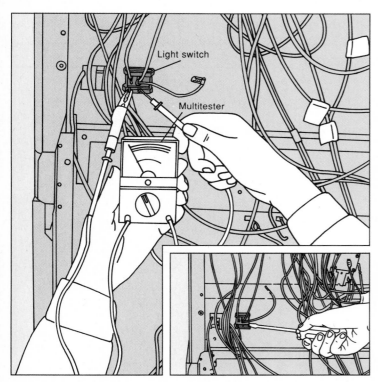

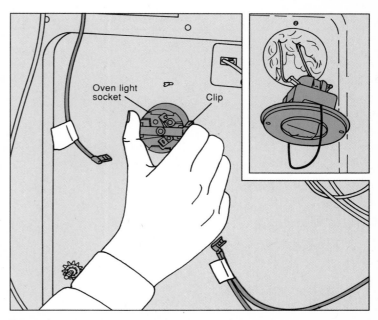

Testing and replacing a panel-mounted light switch. Disconnect power to the range and open the control panel *(page 44)*. Remove one wire from the light switch and clip one miltitester probe to each terminal. With the switch off there should be no continuity. Flip the switch on; the tester should show continuity. To replace a faulty switch, press in the spring clips with a screwdriver *(inset)*. Push the switch out through the front of the range. Disconnect the wires; cut them if they are permanently attached. Connect or splice the wires to the new switch *(page 136)* and snap it into place.

Replacing an oven light socket. Some light sockets are removed through the back of the range. With the power disconnected, remove the rear panel *(page 44)*. The socket's ceramic base has spring clips; use a screwdriver to push in the clips, then pull out the socket, as shown. On some models, you must first unscrew a socket assembly, and then push the socket out of the assembly. Disconnect the wires, insert a new socket and reconnect the wires. In other ranges, the light socket is accessible through the oven. Disconnect the power. Unscrew the socket assembly and pull it into the oven *(inset)*. Disconnect the wires and release the socket by squeezing locking tabs on each side. Snap in a new socket, connect the wires, and replace the assembly.

CHECKING AND REPLACING A RANGE FUSE

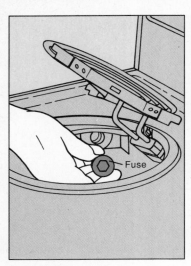
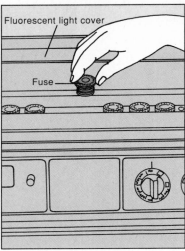
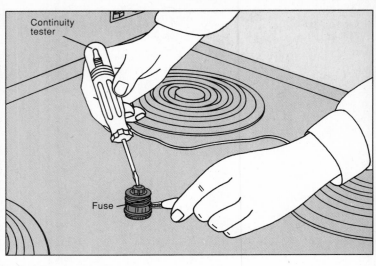

Replacing a range fuse. The fuse, usually found under the left or right front element, protects the lights, clock and appliance receptacle from excessive current. Disconnect power to the range. Pull out the element, or raise it out of the way, and remove the drip pan. Open the fuse cover and unscrew the fuse *(above, left)*. In some ranges, there are fuses above the control panel, under the fluorescent light cover *(above, center)*, or in the top of the storage drawer. Inspect the fuse; a blackened window or broken metal strip means the fuse is blown. If in doubt, test a fuse with a continuity tester *(above, right)*. If the tester does not light, the fuse is blown; insert a new one of the same rating. Close the fuse cover and replace the drip pan and element. Some ranges have small circuit breakers on the control panel instead. Check your owner's manual for the location of fuses.

SERVICING AN APPLIANCE RECEPTACLE

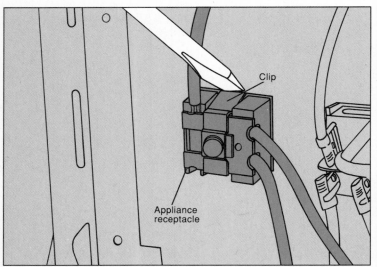

Checking and dismounting the receptacle. The appliance receptacle operates on the 120-volt circuit within the range. If it doesn't work, first check the range fuse. If the receptacle is a timed type, make sure that the timer is properly set; refer to your owner's manual. To check the receptacle, turn off the power and open the control panel *(page 44)*. Inspect the wiring and terminals for burns and corrosion, and tighten loose connections. To remove the receptacle, push in the spring clips on each side with a screwdriver *(above, left)* and pull the receptacle out through the front of the range *(above, right)*. Label and disconnect the wires. Snap in a new receptacle and attach the gray or white wire to the silver-colored terminal, and the black or red wire to the brass-colored terminal. Make sure the green ground wire is connected to the range. If your replacement receptacle has wires already attached, cut the wires to the old receptacle and splice them to the new wires *(page 136)*.

ADJUSTING THE OVEN DOOR

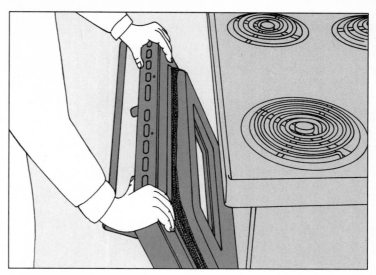

1 Loosening the oven door panels. The oven door can become warped over time, resulting in a poor seal. Discoloration or traces of soot around the door indicate that heat is escaping from the oven. Before adjusting a removable door *(page 44)*, check that it is seated properly on its hinges. To adjust the door, first turn off power to the range. Open the door and loosen—but don't remove—the screws that secure the inner door panel to the outer panel, as shown. You may also have to loosen screws on the door handle and around the edge of the outer panel.

2 Adjusting the door fit. Holding the door at the top, twist it gently from side to side to straighten it. On oven doors with a glass front or a window, be careful to shift the door only slightly. Partially tighten the door screws. Check the seal by pressing the top corners of the door against the oven. You may have to adjust the door several times for a good fit. Tighten the screws securely but do not overtighten—the porcelain could chip.

ADJUSTING THE OVEN DOOR SPRINGS

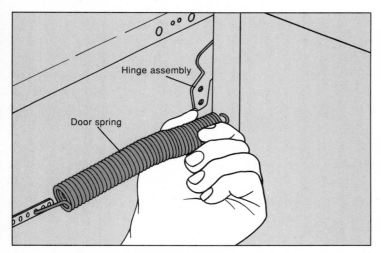

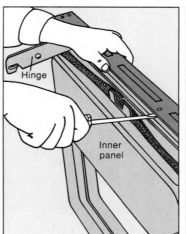

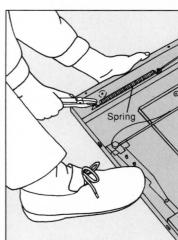

Adjusting cabinet-mounted springs. Turn off power to the range and remove the oven door *(page 44)*. Hinge arms that protrude from the door indicate door-mounted springs, described at right. Otherwise, the springs are cabinet-mounted, as shown. Pull out the lower storage drawer. If you can't find the springs on each side of the cabinet, remove the side panels or call for service. (The springs on wall ovens and slide-in ranges are usually not accessible.) If one spring is broken, replace both. Wearing safety goggles, lift and unhook the hinge end, as shown; if the spring is very stiff, grasp it with locking pliers. To increase the door tension, rehook the spring into a lower hole on the hinge assembly; to decrease tension hook it into a higher hole. Repeat with the other spring. Replace the door; if the tension is incorrect, readjust the springs.

Adjusting door-mounted springs. Remove the oven door *(page 44)*. Remove the screws on the inner door panel and along the edges of the outer panel. You may also have to remove the door handle. There may be tabs on one panel that fit into slots in the other; use a screwdriver to pry up the tabs *(above, left)*. Starting at the top, lift the inner panel free. Remove any insulation covering the springs. Lay the door on a table or the floor and inspect the springs; if a spring is worn or broken, replace both. Wearing safety goggles, unhook one end of the spring and insert it into the next hole in the hinge assembly. If the spring is stiff, brace the door with your hand or foot and use locking pliers to pry up one end of the spring and rehook it firmly in place *(above, right)*. Reassemble the door.

REPLACING AN OVEN GASKET (Cabinet-mounted ovens)

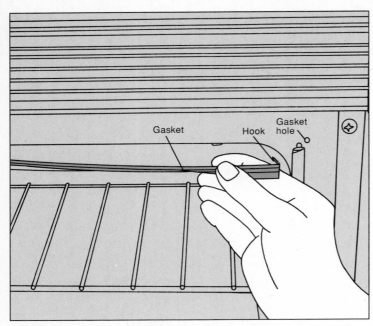

1 Checking the oven gasket. The gasket on many ovens is a simple rubber channel clipped onto the cabinet. To replace this gasket, disengage the damaged section by hand and hook a new one in place, as shown. Other cabinet-mounted gaskets are clamped between the oven liner and the range cabinet (step 2).

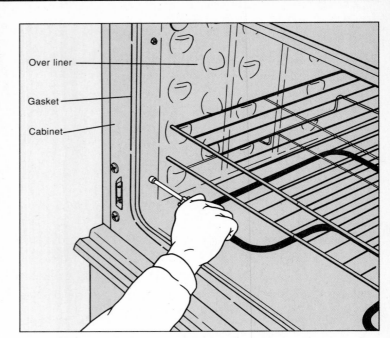

2 Removing the oven retaining screws. To remove a gasket clamped between the oven liner and the range cabinet, you must free the oven liner and pull it forward. Disconnect power to the range and remove the oven door to make the job easier. Check around the front edge of the oven for screws or clips holding the oven liner in place; remove them as above and proceed to step 4. If there are no screws, the oven liner is probably fastened at the back (step 3).

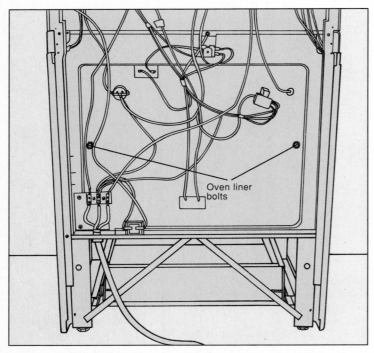

3 Loosening the oven liner bolts. Disconnect power to the range and pull it away from the wall. Unscrew the back panel (page 44). Check for two bolts that protrude from the back of the range, one at each side (above). If you can't find them, call a service technician to replace the gasket. Loosen the nuts on the bolts about a quarter of an inch.

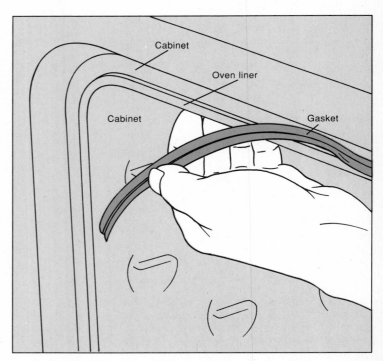

4 Removing and replacing the gasket. Pull out the oven liner slightly by rocking it back and forth. Disengage the gasket from between the liner and cabinet, as shown. Position the lip of a new gasket behind the rim of the oven liner. Push the oven liner back into place and either replace the screws at the front or tighten the nuts at the back. Replace the oven door or rear panel.

REPLACING A DOOR-MOUNTED GASKET (Self-cleaning ovens)

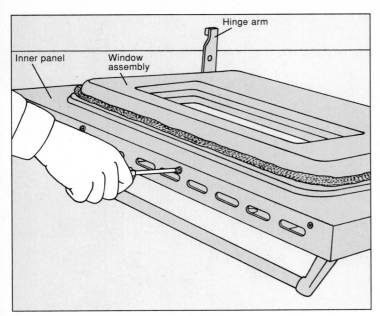

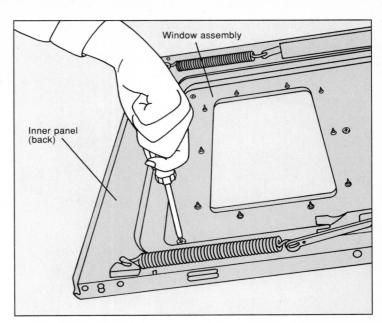

1 **Separating the door panels.** On self-cleaning ovens, the gasket is held between the panels of the oven door and can be removed only by disassembling the door. Turn off power to the range, unscrew the door-hinge arms and take off the door. Remove the screws on the inner panel and along the outer edges of the door, as shown. You may also have to remove the door handle. There may be tabs on the outer panel that fit into slots in the inner panel; use a screwdriver to pry them apart if necessary. Starting at the top, lift off the inner panel and window assembly.

2 **Removing the door window.** Position the inner door panel with the gasket facing down. Remove the screws that hold the window assembly to the panel, as shown. On some models, you must first remove a metal window shield and a layer of insulation to reach the window assembly. Lift off the window assembly to reveal the gasket attachment.

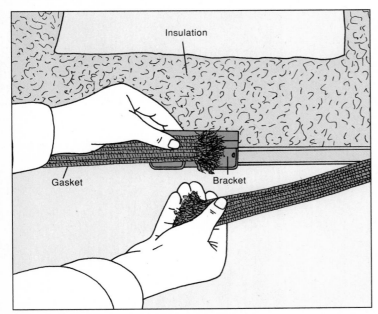

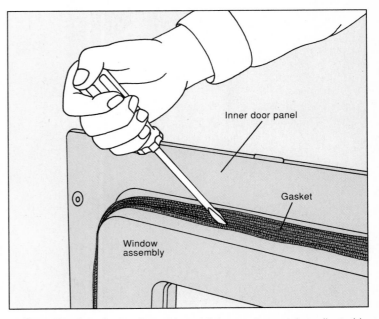

3 **Removing and replacing the gasket.** The gasket fits between the window assembly and the inner door panel. If it is held by clips, unscrew them and slip out the gasket. Install a new gasket under the clips and tighten the screws. If the gasket is held between the window assembly and the door panel, just lift the gasket off (you may first have to unhook it from a bracket). Position the new gasket with the small bead, or edge, against the panel edge and hook the two ends to the bracket, as shown. Screw the window assembly and the door panel together loosely.

4 **Adjusting the gasket.** A loose-fitting gasket can be adjusted by using the flat edge of a screwdriver to wedge the excess between the door panel and window assembly. Starting at the top of the door, push the gasket in with the screwdriver, as shown, tightening the screws together as you go. Replace the insulation and refasten the window shield if the door has one. Screw the inner and outer door panels back together and replace the door on its hinges. If necessary, adjust the fit of the oven door *(page 55)*.

SERVICING THE RANGE HOOD (Fan-type)

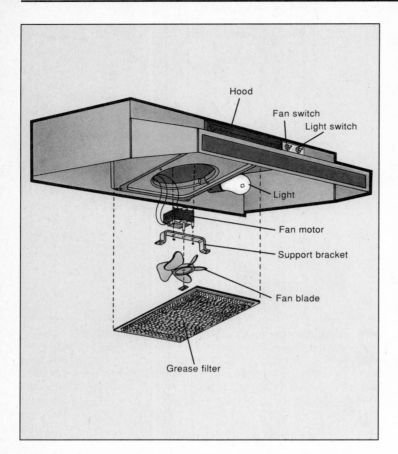

Hood
Fan switch
Light switch
Light
Fan motor
Support bracket
Fan blade
Grease filter

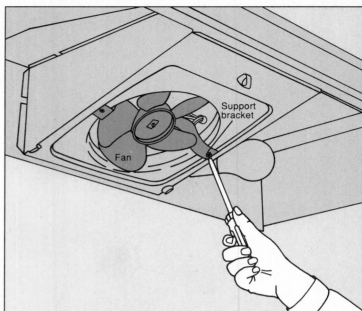

Support bracket
Fan

1 Removing the fan. Disconnect power to the range hood. To get at the fan, first remove the aluminum grease filter, which is secured by clips. Wash the filter in hot, soapy water or in the dishwasher. Do this weekly or after cooking any greasy food; replace the filter annually. To release the fan, remove the screws that hold the support bracket to the housing.

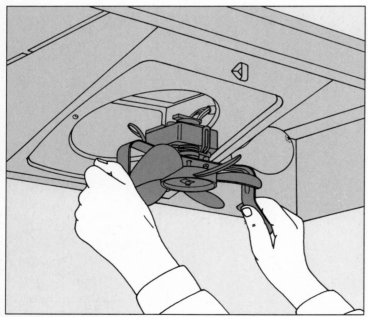

2 Cleaning the fan. Holding the fan assembly by its support bracket, lower it from the housing, as shown. Wash the fan blades with a soapy cloth and dry them thoroughly. Wipe the motor with a dry cloth. With a soapy cloth, wipe the interior of the hood and as much of the ductwork as you can reach to remove grease and dirt. At the same time, check for any obstructions which may be hindering the operation of the fan.

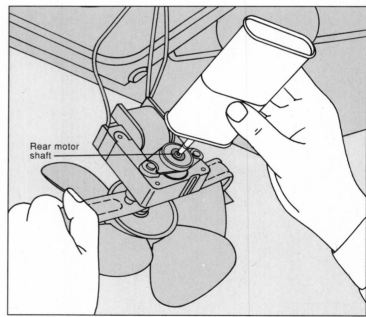

Rear motor shaft

3 Oiling the motor shaft. Even for permanently lubricated motors, a few drops of machine oil will improve efficiency. Hold the fan by the support bracket. Put a few drops of oil on the rear motor shaft. To reassemble, position the fan assembly in the housing, screw the support bracket in place and slide in the grease filter.

RANGE HOOD (Squirrel-cage)

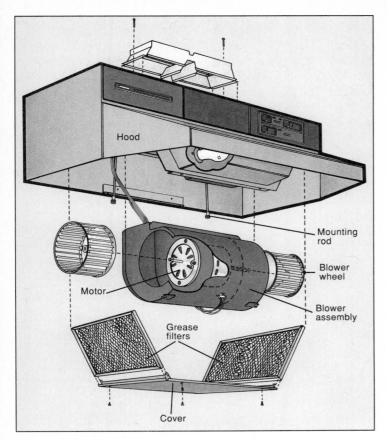

Hood

Motor

Mounting rod

Blower wheel

Blower assembly

Grease filters

Cover

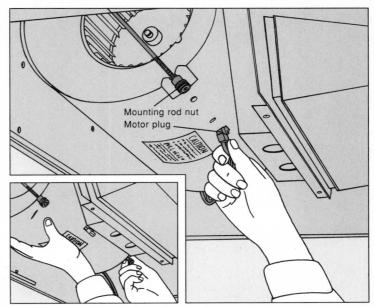

Mounting rod nut
Motor plug

1 **Removing the grease filters and blower assembly.** Disconnect power to the range hood. Pull out the aluminum grease filters and wash them out in hot, soapy water or in the dishwasher. Remove the screws at each side of the cover and remove it to expose the blower assembly. Unplug the motor *(above)*. Supporting the blower assembly with one hand, loosen, but do not remove, the mounting rod nuts on each side *(inset)*. Move the rods out of the brackets and lower the blower assembly.

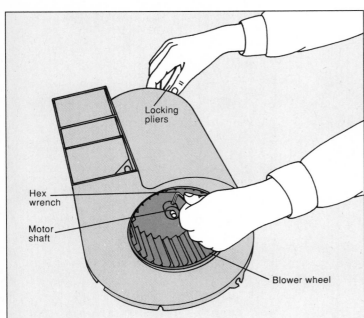

Locking pliers

Hex wrench

Motor shaft

Blower wheel

2 **Servicing the blower wheels.** Using a hex wrench, remove the setscrews holding the blower wheels to the motor shaft. Grip the other end of the shaft with locking pliers, the jaws covered with masking tape to protect the shaft. Slide the wheels off the shaft, handling them carefully; the aluminum blades are easily damaged. Wash the wheels in hot, soapy water or in the dishwasher. Replace them on the motor shaft and tighten the setscrews.

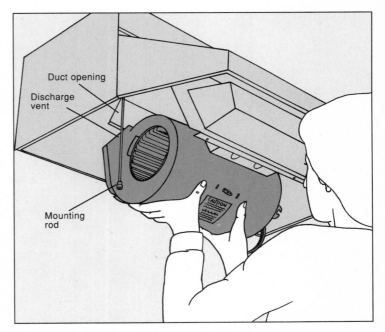

Duct opening

Discharge vent

Mounting rod

3 **Reinstalling the blower assembly.** First use a soapy cloth to clean the interior of the range hood and as much of the ductwork as you can reach. Lift the blower assembly into position under the range hood *(above)*. Be sure to align the discharge vent of the assembly with the duct opening. Slip the mounting rods into the brackets and tighten the nuts by hand. Plug in the motor. Screw the blower assembly cover in place and reinsert the filters.

GAS RANGES

A gas range has few moving parts, and many older machines have no electrical parts at all. The two basic types of gas ranges differ mainly in the way they ignite the gas. In older models, gas flows to the range and oven burners and is ignited along the way by a small pilot light. Ranges with electric igniters use sparks or an electrically heated coil called a glowbar to ignite the gas. The igniters are wired to an ignition module on the back of the range that produces the high voltage required for sparking. On both types, air shutters control the amount of air mixed with the gas flowing to the burner. A thermostat regulates oven temperature by turning the gas on and off.

Most gas ranges have similar components, and the repairs on the following pages can be adapted to older and newer models alike. Most problems are caused by dirt—spilled food often clogs burner portholes and grease can block air shutters. Clean often under the cooktop and wash the burners as needed.

Repairs that involve the gas supply line, such as replacing a burner control or a thermostat, carry the risk of creating gas leaks and should always be left to a professional. If the supply line is a rigid pipe, never move the range; have the gas company or a service technician disconnect and move it for you. Make sure you know the location of the gas shutoff valve for your range, and how to turn it off; it is usually on the supply pipe behind the range *(page 140)*.

Some gas range repairs—replacing a clock or gasket, or adjusting an oven door—are the same as for electric ranges. If a repair is not listed here, consult the electric range Troubleshooting Guide *(page 43)*.

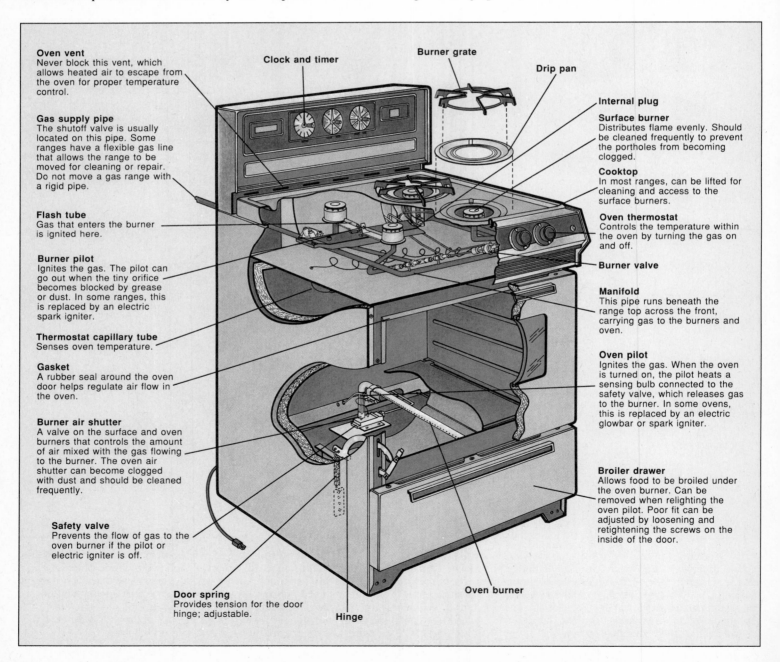

Oven vent
Never block this vent, which allows heated air to escape from the oven for proper temperature control.

Gas supply pipe
The shutoff valve is usually located on this pipe. Some ranges have a flexible gas line that allows the range to be moved for cleaning or repair. Do not move a gas range with a rigid pipe.

Flash tube
Gas that enters the burner is ignited here.

Burner pilot
Ignites the gas. The pilot can go out when the tiny orifice becomes blocked by grease or dust. In some ranges, this is replaced by an electric spark igniter.

Thermostat capillary tube
Senses oven temperature.

Gasket
A rubber seal around the oven door helps regulate air flow in the oven.

Burner air shutter
A valve on the surface and oven burners that controls the amount of air mixed with the gas flowing to the burner. The oven air shutter can become clogged with dust and should be cleaned frequently.

Safety valve
Prevents the flow of gas to the oven burner if the pilot or electric igniter is off.

Door spring
Provides tension for the door hinge; adjustable.

Hinge

Clock and timer

Burner grate

Drip pan

Internal plug

Surface burner
Distributes flame evenly. Should be cleaned frequently to prevent the portholes from becoming clogged.

Cooktop
In most ranges, can be lifted for cleaning and access to the surface burners.

Oven thermostat
Controls the temperature within the oven by turning the gas on and off.

Burner valve

Manifold
This pipe runs beneath the range top across the front, carrying gas to the burners and oven.

Oven pilot
Ignites the gas. When the oven is turned on, the pilot heats a sensing bulb connected to the safety valve, which releases gas to the burner. In some ovens, this is replaced by an electric glowbar or spark igniter.

Broiler drawer
Allows food to be broiled under the oven burner. Can be removed when relighting the oven pilot. Poor fit can be adjusted by loosening and retightening the screws on the inside of the door.

Oven burner

TROUBLESHOOTING GUIDE

SYMPTOM	POSSIBLE CAUSE	PROCEDURE
Gas odor	Pilot light out	Ventilate room; relight pilot *(pp. 63, 66)* □○
Gas odor with all pilots lit or with electric ignition	Burner control on slightly	Turn off burner controls
	Gas line leak	Turn off gas to range, ventilate room, call gas company *(p. 12)*
Surface burner doesn't light	Pilot light out	Relight pilot *(p. 63)* □○
	Burner pilot porthole blocked	Clear burner portholes *(p. 64)* □○
	Burner or flash tube out of position	Reposition burner or flash tube *(p. 64)* □○
	No power to range (ranges with electric igniters)	Check that range is plugged in; check for blown fuse or tripped circuit breaker *(p. 132)* □○
	Electric igniter doesn't spark	Inspect igniter and ignition module *(p. 65)* ◨◓
	Burner flame openings clogged	Clean burner *(p. 64)* □○
	Too much air to burner	Adjust air shutter *(p. 64)* □○
	Not enough gas to burner	Call for service
Surface burner pilot doesn't stay lit	Pilot opening blocked	Clear pilot opening *(p. 63)* □○
	Pilot light too low	Adjust pilot light *(p. 63)* □○
	Too much air to burner	Adjust air shutter *(p. 64)* □○
	Not enough gas to burner	Call for service
Surface burner flame low or uneven	Burner flame openings clogged	Clean burner *(p. 64)* □○
	Not enough air to burner	Adjust air shutter *(p. 64)* □○
	Not enough gas to burner	Call for service
Surface burner flame too high, noisy or blowing	Too much air to burner	Adjust air shutter *(p. 64)* □○
	Too much gas to burner	Call for service
Surface burner flame yellow or sooty	Burner out of position	Reposition burner *(p. 64)* □○
	Not enough air to burner	Adjust air shutter *(p. 64)* □○
	Too much gas to burner	Call for service
Oven burner doesn't light	Pilot light out	Relight pilot *(p. 66)* □○
	Pilot light too low	Adjust pilot light *(p. 66)* □○
	Clock timer set improperly	Check Use and Care manual; reset timer
	No power to range (ranges with electric igniters)	Check that range is plugged in; check for blown fuse or tripped circuit breaker *(p. 132)* □○
	Electric igniter doesn't spark	Inspect igniter *(p. 68)* □○ and ignition module *(p. 65)* ◨◓
	Fuses blown (ranges with glowbar igniter)	Test fuse *(p. 69)*
	Glowbar igniter faulty	Check glowbar igniter *(p. 69)* ◨◓
	Flame switch faulty	Test flame switch *(p. 68)* ◨◓
	Thermostat, selector switch or safety valve faulty	Call for service
Oven burner pilot doesn't stay lit	Pilot light too low	Adjust pilot light *(p. 66)* □○
	Pilot opening blocked	Clear pilot opening *(p. 63)* □○
Oven doesn't hold set temperature; oven bakes unevenly	Door not aligned or gasket faulty	Check door *(p. 55)* □○ and oven gasket *(pp. 56, 57)* □○
	Burner flame openings clogged	Clear burner openings *(p. 67)* □○
	Thermostat or capillary tube faulty	Call for service
Self-cleaning oven doesn't clean	Control setting incorrect	Check Use and Care manual; reset controls
	Oven door not locked	Reclose and lock door
	Thermostat, selector switch or door lock faulty	Call for service
	Door not aligned or gasket faulty	Check door *(p. 55)* □○ and oven gasket *(pp. 56, 57)* □○
Electrical accessory doesn't work	No power to range	Check that range is plugged in; check for blown fuse or tripped circuit breaker *(p. 132)* □○
	Internal range plug loose or faulty	Inspect internal range plug *(p. 62)* ◨◓
	Electrical accessory faulty	See electric range Troubleshooting Guide *(p. 43)*

DEGREE OF DIFFICULTY: □ **Easy** ◨ **Moderate** ■ **Complex**
ESTIMATED TIME: ○ **Less than 1 hour** ◓ **1 to 3 hours** ● **Over 3 hours**

ACCESS TO THE RANGE

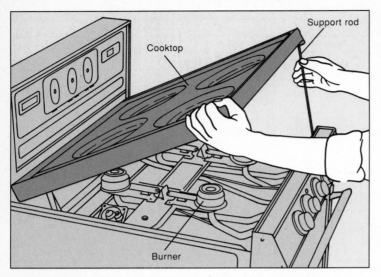

Opening the cooktop. To reach the surface burners and pilots or electric igniters, remove the burner grates, grasp the front edge of the cooktop or the two front burner wells, and lift. Prop the cooktop open with the support rod inside. To replace the cooktop, lift it off the support rod and lower it gently; lowering it too rapidly can blow out the pilot lights. On some ranges, the cooktop can be removed completely by lifting it and pulling it toward you.

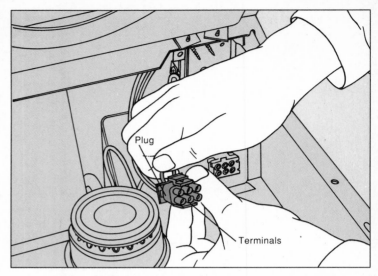

Disconnecting an internal range plug. In gas ranges with electrical components—clock, lights, ignition system—an internal harness plug connects the power cord to the internal wiring. You can disconnect power to the range by unplugging this assembly. Raise the cooktop *(left)*, locate the plug (usually at the right rear of the machine) and pull to disconnect it. Examine the plug for burned or pushed-in terminals *(above)*. Reposition the terminals by hand if you can, and reconnect the plug. If the components still don't work, or if the terminals are burned, replace the plug: Cut the wires, splice on the wires of a replacement plug *(page 136)* and reconnect the plug.

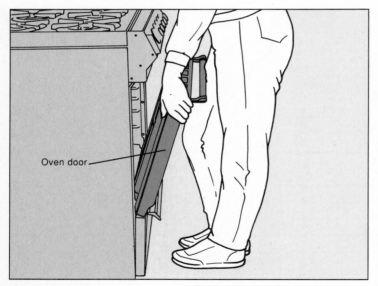

Removing the oven door. Most newer ranges have a removable oven door, simplifying access to the oven burner assembly and pilot light or electric igniter. To remove the door, open it to the first stop, the broil position, and grip each side. Supporting the door with your knee, pull it up straight off its hinges. The spring-mounted hinge arms can slam shut against the range if accidentally touched; you may close them against the cabinet for safety.

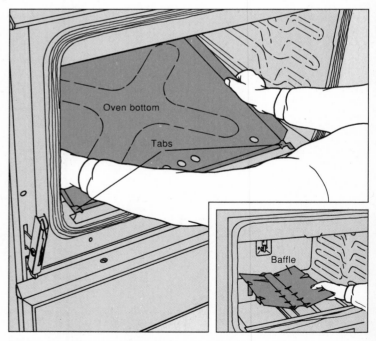

Removing the oven bottom and baffle. On many ranges you can simply lift out the oven bottom. For some models, you must first slide forward small locking tabs at the front or rear of the oven bottom. If screws hold the oven bottom in place, loosen them slightly and slide them back to release the bottom. To replace a broken or corroded oven bottom, reverse these steps. To remove the oven burner baffle beneath the oven bottom, remove the wing nut or screws holding it in place. Then lift it up and out *(inset)*.

LIGHTING AND ADJUSTING SURFACE BURNER PILOTS

Relighting a surface burner pilot.
Caution: If the pilot has been out for some time, or you detect a strong odor of gas, ventilate the room and call for service *(page 12)*. Otherwise, turn off all range controls and prop open the cooktop *(page 62)*. Place a lighted match near the opening of the pilot, located midway between two burners, as shown. If the pilot does not stay lit, clean it *(below, left)* or adjust it *(below, right)*.

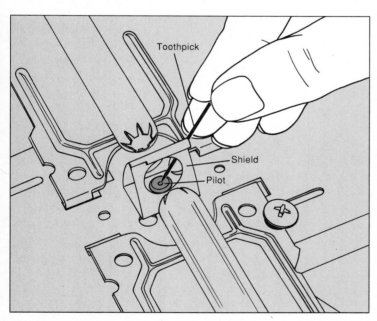

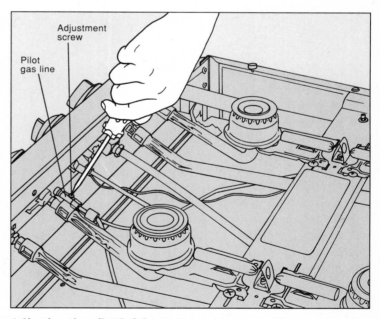

Cleaning a pilot. The small opening of a pilot can easily become clogged with dust, grease or food, preventing proper gas flow to the pilot light. A pilot that won't light, or won't stay lit, may simply need cleaning. Insert a sharp wooden toothpick in the pilot opening, as shown, and move it up and down gently, taking care not to enlarge or deform the opening. If a protective metal shield over the pilot prevents you from reaching the opening, remove it by pressing in the tabs on either side. Clean an oven burner pilot the same way.

Adjusting the pilot height. A pilot that frequently blows out may be set too low or too high. To adjust it, turn off the controls and prop open the cooktop *(page 62)*. The pilot adjustment screw is usually on the side of the pilot, or on the pilot gas line near the manifold at the front of the range. (On some ranges, the adjustment screw is behind the burner control knob. Pull off the knob and insert a screwdriver into the opening beside the stem to adjust the pilot.) Using a screwdriver, turn the screw slowly counterclockwise to increase the size of the pilot light, as shown. The flame should be a sharp blue cone, 1/4 to 3/8 inch in height. Replace the cooktop when the pilot is properly adjusted.

CLEANING AND ADJUSTING SURFACE BURNERS

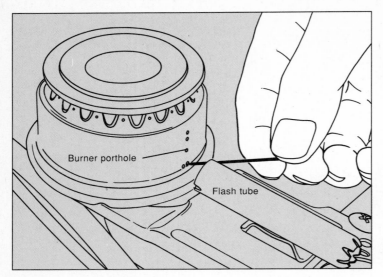

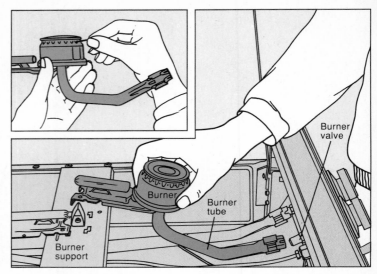

Cleaning the burner portholes. If a burner will not light, raise the cooktop *(page 62)* and check that the pilot light *(page 63)* or spark igniter *(page 65)* is working. Also, check that the burner is properly seated on its support—the flash tube should form a straight line from the pilot or spark igniter to the burner portholes. If the portholes are clogged, gas cannot flow into the flash tube for ignition. Use a needle or fine wire—or even a straightened paper clip—to ream out the burner portholes opposite the flash tube, as shown.

Removing a surface burner. On most ranges, the surface burners are designed to be removed easily for cleaning. On some older ranges, only the top ring of the burner comes off; the burner base remains in place. Prop open the cooktop *(page 62)*. Lift the burner from its support, then pull the burner tube backward off the burner valve, as shown. If the surface burners were screwed to the supports for shipping, discard the screws; they can cause the burners to warp out of alignment with the flash tubes. Wash the burner in hot, soapy water, and scrub the portholes with a brush. After washing, use a pin or wire to clean the flame openings *(inset)*. Let the burner dry thoroughly. To replace the burner, insert the tube onto the burner valve and lower the burner onto its support.

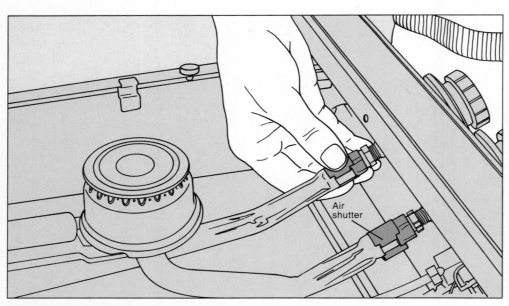

Adjusting a surface burner air shutter. A shutter or sleeve on the burner tube controls the amount of air that mixes with the gas flowing to the burner. An improper gas/air mixture results in an inefficient flame *(right)*. To adjust the air shutter, turn off all controls and raise the cooktop *(page 62)*. Locate the air shutter—on the burner tube near the manifold at the front of the range—and loosen the screw holding it in place. Turn the burner control on HIGH and twist or slide the shutter open, as shown, until the flame is fed excessive air. Slowly close the shutter until the flame is the correct size and color. Turn off the burner control, tighten the shutter screw and replace the cooktop. On models with an air-mixing chamber, loosen the retaining screw, slide the plate to adjust the air intake and tighten the screw.

Insufficient air
Not enough air results in a weak flame without sharp blue cones. The flame may be red, yellow or yellow-tipped, and may leave soot deposits on pots and pans.

Excessive air
An unsteady, blowing flame is a sign of too much air. The flame may not burn all around and may be noisy. Too much air can also prevent the burner from igniting.

Correct air adjustment
A properly adjusted burner has a steady, quiet flame with sharply defined blue cones about 1/2 to 3/4 inch in height.

SERVICING THE ELECTRIC IGNITER

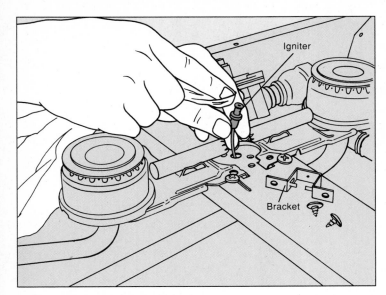

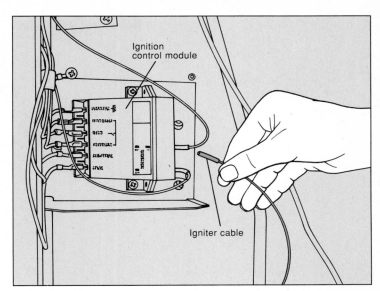

1 **Checking the electric igniter.** If a burner won't light, raise the cooktop *(page 62)* and observe the igniter as you turn on the burner. If it doesn't spark, turn on the second burner; if the igniter now works, the first burner control is faulty; call for service. If the second burner doesn't work, the igniter or the ignition control module may be at fault. Unscrew the bracket over the igniter and inspect it; replace the igniter if cracked or burned *(step 2)*. If the igniter is dirty, clean it gently with a cotton swab or clean rag, as shown, and check it again. Replace the ignition control module if no igniters work *(step 4)*.

2 **Disconnecting the electric igniter.** Electric igniters are wired to an ignition control module that produces high voltage for the spark. Turn off power to the range by disconnecting the plug under the cooktop *(page 62)*, or by unplugging the range from the outlet. (**Caution:** If the gas supply line is a rigid pipe, do not attempt to move the range.) Pull the range away from the wall and trace the cable from the igniter to the module, a plastic or metal box mounted on the back of the range (in some models, it is underneath the oven bottom). With a screwdriver, remove the metal plate covering the module. Pull the igniter cable from the module terminal, as shown.

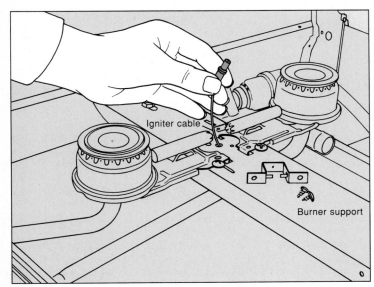

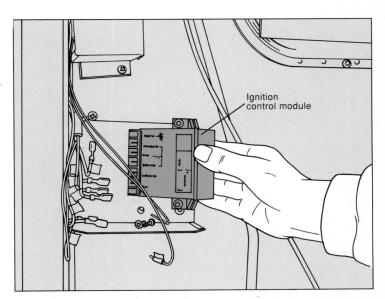

3 **Replacing the igniter.** Unscrew the bracket over the igniter. The cable may also be fastened with clips to the underside of the burner support; remove the clips. Pull the igniter cable through the hole in the burner support, as shown. To simplify threading the new cable, tie a string to the end of the old one and pull it through the range as you pull out the cable. Tie the string to the end of the new cable and pull it back through to the module. Seat the igniter firmly on the burner support. Connect the new cable to the ignition module and replace the module cover.

4 **Replacing the ignition control module.** Locate and uncover the ignition control module *(step 2)*. Use masking tape to label the positions of all the wires leading to the module, then disconnect them. Unscrew the module from the range. To install a new module, first screw it to the range, then reconnect the wires and replace the cover plate. Push the range back against the wall, turn on the power and check the operation of the burners.

LIGHTING AND ADJUSTING THE OVEN PILOT

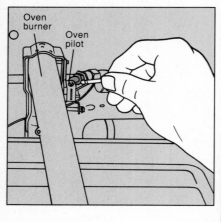

Oven burner
Oven pilot

Relighting the oven pilot. Turn off all the range controls. If your range is equipped with a broiler drawer, open it and wait five minutes for gas to dissipate. To light the pilot, reach back with a lighted match to the tip of the pilot on the burner assembly *(near left)*. Otherwise, open the oven door and remove the oven bottom and baffle *(page 62)*. Place a lighted match near the tip of the pilot on the burner assembly *(far left)*. On some older ranges, you must hold down a button on the side of the oven or on the oven thermostat as you light the pilot. Replace the baffle and oven bottom. Turn on the oven and wait a minute or two for the burner to light; if it doesn't, adjust the pilot *(next step)*.

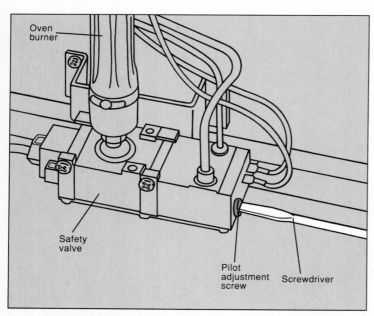

Oven burner

Safety valve

Pilot adjustment screw   Screwdriver

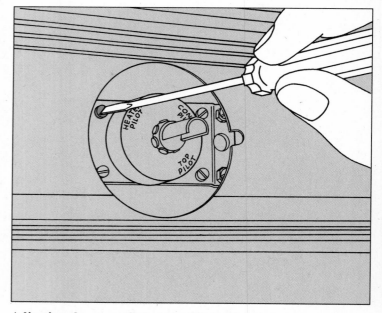

Adjusting the oven pilot. If the oven pilot flame is set too low, the burner will not light even if the pilot is on. To adjust the pilot, turn off all controls and remove the oven bottom and baffle *(page 62)*. Examine the safety valve on the burner assembly at the rear of the oven for a slotted screw near the gas line to the pilot. (If the adjustment screw is not on the safety valve, proceed to the next step). With a screwdriver, turn the screw counterclockwise in tiny increments, as shown, until the pilot has a steady, blue, yellow-tipped flame about 1/4-inch high. If the pilot will not light, turn the screw slightly, then light the pilot. Turn on the oven thermostat and observe the burner and pilot. If the pilot flame does not increase, the thermostat is probably defective. If the burner does not light within four minutes, the safety valve must be replaced. In either case, call for service.

Adjusting the oven pilot at the thermostat. Pull off the oven temperature control knob and check for a slotted screw (sometimes marked "Pilot" or "P"). If you can't find the screw at the front of the thermostat, raise the cooktop *(page 62)* and inspect the back of the thermostat for a slotted screw near the gas line to the oven pilot. Turn off all controls, open the oven door and remove the oven bottom and baffle *(page 62)*. Adjust the pilot flame as described in the previous step; check for the same results and call for service if necessary. Gently push the control knob back on its shaft or lower the cooktop.

CLEANING AND ADJUSTING THE OVEN BURNER

Wire

Burner

Cleaning the oven burner. Poor baking results, or an odor of gas when the oven is on, may be due to the uneven flame produced by a clogged burner. With all controls off, remove the oven door and take out the oven bottom and baffle *(page 62)*. Turn on the oven and observe the burner. If the flame is not continuous along the length of the burner, some of its holes may be clogged. Turn off the oven control. Poke a thin, stiff wire or a needle into each hole in the sides of the burner, as shown, and rotate it to clear the blockage.

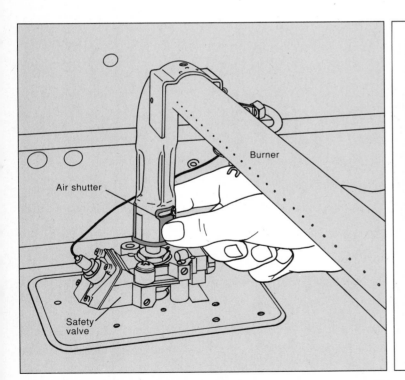

Burner

Air shutter

Safety valve

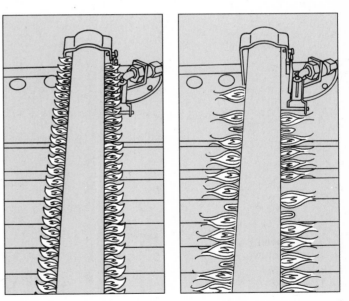

Adjust the oven flame to a steady, blue, one-inch cone, with a distinct inner cone of about 1/2 inch *(left)*. Too much air will cause a hissing flame *(right)*; with too little air, the flame will be yellow-orange.

Adjusting the oven burner flame. With all controls off, remove the oven door, oven bottom and baffle *(page 62)*. Locate the air shutter at the base of the oven burner, just above the safety valve. Loosen the shutter screw and rotate the air shutter to increase or decrease the shutter opening *(above, left)*; the larger the air opening, the larger the flame. Some older ranges have an air-mixing chamber with a sliding plate that controls the size of the air opening; loosen the screw and slide the plate. Remove your hand from the oven and turn on the oven temperature control to observe the flame. If it is not the proper size *(above, right)* and color, turn off the oven and readjust the air shutter. When the flame is correct, tighten the shutter screw and replace the baffle and oven bottom.

CHECKING AND REPLACING THE OVEN FLAME SWITCH

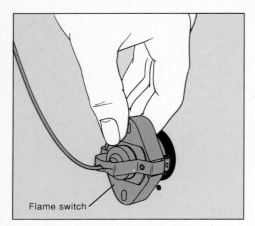

Flame switch

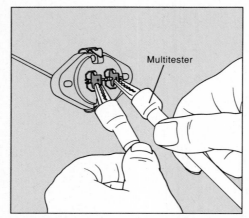

Multitester

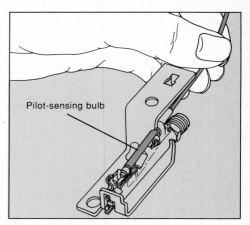

Pilot-sensing bulb

1 **Removing the flame switch.** In some gas ranges, an oven flame switch prevents gas flow to the burner if the pilot goes out. If the pilot is lit but the oven burner doesn't ignite, the problem may be a faulty flame switch. Disconnect the plug at the rear *(page 62)*. Remove the oven door, the oven bottom and the baffle *(page 62)*. Remove the screws holding the flame switch to the oven wall and pull it forward, as shown.

2 **Testing the flame switch.** Disconnect the two wires from the flame switch terminals at the back, but do not remove the wire in the front that leads to the sensing bulb. Clip one probe of a continuity tester or a multitester set at RX1 to each terminal and test for continuity. If the tester does not show continuity with the oven pilot lit, replace the switch. If the switch has continuity, the oven thermostat is probably faulty; call for service.

3 **Replacing the flame switch.** To remove a defective flame switch, gently work the pilot-sensing bulb loose from its bracket near the pilot. The bulb may contain mercury; take care not to break it. To install a new flame switch, replace the pilot-sensing bulb in its bracket, reconnect the two wires to the new switch and screw the switch to the oven wall. Replace the oven bottom, baffle and door.

CHECKING AND REPLACING THE OVEN ELECTRIC IGNITER

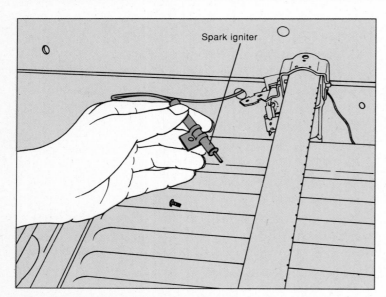

Spark igniter

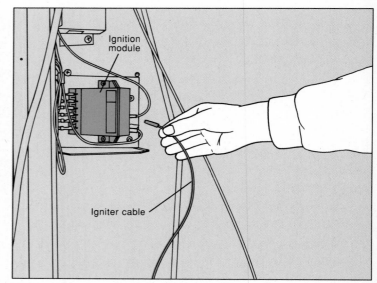

Ignition module

Igniter cable

1 **Checking the oven spark igniter.** When the oven thermostat is turned on, the igniter spark lights the pilot flame, which heats a sensing bulb connected to the safety valve. At a certain temperature, the safety valve opens and gas flows to the burner, where it is ignited. If your oven won't light, first check the igniter. Turn off power to the range and remove the oven door, bottom and baffle *(page 62)*. Unscrew the igniter from its mounting bracket and inspect it for cracks or other flaws, as shown. Replace the igniter if damaged.

2 **Replacing the oven spark igniter.** The igniter is connected to the ignition control module, which is usually mounted on the back of the range; in some models it is under the oven bottom. Pull the range away from the wall. (**Caution**: Do not move the range if its gas supply line is a rigid pipe.) Trace the igniter cable to the module and unscrew its cover plate. To help you thread the cable, use a string *(page 65)*. Disconnect the igniter cable from the module and pull the igniter and cable out through the oven. Screw the new igniter to the bracket with its electrode 1/4 inch from the pilot. Thread the cable through the back of the oven and connect it to the ignition module.

TESTING AND REPLACING THE OVEN GLOWBAR IGNITER

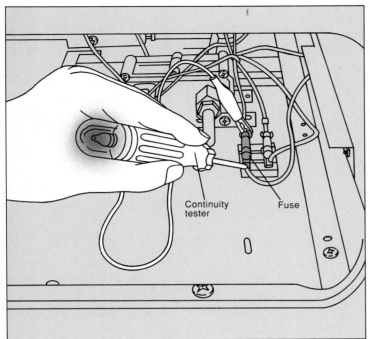

Continuity tester   Fuse

1 Testing the fuse. If your oven does not heat, check its cartridge fuse with a continuity tester or a multitester set at RX1. Raise the cooktop and disconnect the internal range plug *(page 62)*. The fuse is near the safety valve, in a metal cover or fuse holder; reach it by removing the broiler drawer or the oven bottom *(page 62)*. (For a self-cleaning range, unscrew the access panel below the oven door and the metal heat shield inside.) Unscrew the fuse cover or the top of the fuse holder. Disconnect one of the wires to the fuse and touch a tester probe to each terminal, as shown. If the fuse is difficult to test in place, use a plastic fuse puller to remove it for testing. If the tester does not show continuity, replace the fuse with one of the same amperage and screw the metal fuse cover back in place.

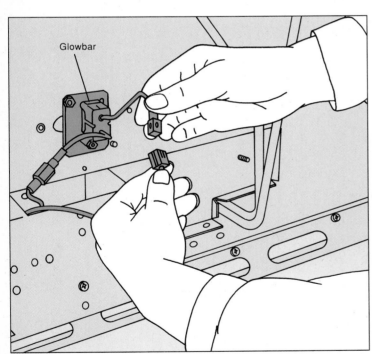

Glowbar

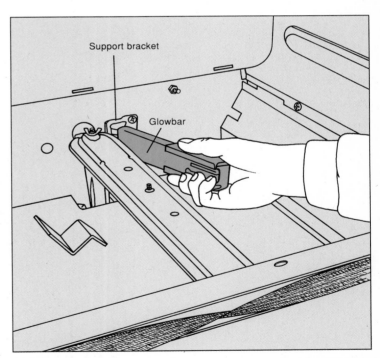

Support bracket   Glowbar

2 Checking the glowbar. First make sure the clock timer is set properly. Turn on the thermostat; if the indicator light does not come on, call for service. To check the glowbar, turn off power to the range and pull it away from the wall. (**Caution:** Do not move your range if its gas supply line is a rigid pipe.) At the back of the range, unscrew the cover plate over the glowbar wiring and disconnect the plugs, as shown. In some ranges, the glowbar wiring is under a metal cover near the oven burner assembly; in this case, remove the oven door and the oven bottom and baffle *(page 62)*. Insert a continuity tester probe into each glowbar terminal; if the tester does not show continuity, replace the glowbar. If the glowbar has continuity, the safety valve may be faulty; call for service.

3 Replacing the oven glowbar. Unscrew the glowbar from the burner support bracket and the oven wall, and pull the glowbar free, as shown. Mount a new glowbar in the burner support bracket at the back of the range, reconnect its plugs at the back of the range and replace the metal cover over the glowbar wiring.

DISHWASHERS

The dishwasher combines water pressure, detergent and heat to clean dishes and kitchen utensils more thoroughly and efficiently than hand washing. Although a dishwasher is a complex machine, its most common problems are usually due to simple failures that are easy to fix.

During a typical 75-minute cycle, the dishwasher tub fills with water, which mixes with detergent released by the detergent dispenser. The detergent-and-water solution is then warmed to about 150°F by the heating element and pumped through the spray tower and spray arm, which spins about 40 times a minute, hurling the mixture against the dishes and washing away even hardened food waste. After the dishes are rinsed and the dishwasher drained, the air inside the machine is warmed by the heating element, drying the dishes. (Some newer models have an energy-saver feature that turns off the heating element during the drying cycle, using a small blower to air-dry the dishes.)

Built-in dishwashers are installed under a kitchen counter with permanent plumbing and wiring connections. Portable models have a plastic coupler so that they can be connected to the sink faucet and drain, casters that allow them to be rolled to and from the sink, and a power cord that plugs into a grounded, 120-volt outlet.

The most common dishwasher problem is incomplete cleaning. Before repairing the appliance itself, check for other possible causes: improper loading, low water temperature, low water pressure, and ineffective detergent. Always rinse dishes, pots and pans in the sink before loading. Large pieces of food can

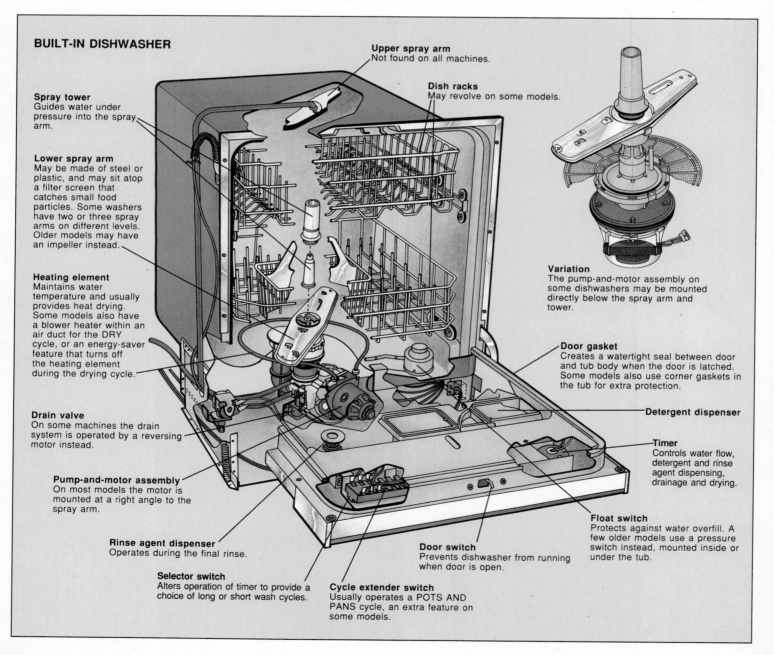

BUILT-IN DISHWASHER

Spray tower
Guides water under pressure into the spray arm.

Lower spray arm
May be made of steel or plastic, and may sit atop a filter screen that catches small food particles. Some washers have two or three spray arms on different levels. Older models may have an impeller instead.

Heating element
Maintains water temperature and usually provides heat drying. Some models also have a blower heater within an air duct for the DRY cycle, or an energy-saver feature that turns off the heating element during the drying cycle.

Drain valve
On some machines the drain system is operated by a reversing motor instead.

Pump-and-motor assembly
On most models the motor is mounted at a right angle to the spray arm.

Rinse agent dispenser
Operates during the final rinse.

Selector switch
Alters operation of timer to provide a choice of long or short wash cycles.

Upper spray arm
Not found on all machines.

Dish racks
May revolve on some models.

Variation
The pump-and-motor assembly on some dishwashers may be mounted directly below the spray arm and tower.

Door gasket
Creates a watertight seal between door and tub body when the door is latched. Some models also use corner gaskets in the tub for extra protection.

Detergent dispenser

Timer
Controls water flow, detergent and rinse agent dispensing, drainage and drying.

Float switch
Protects against water overfill. A few older models use a pressure switch instead, mounted inside or under the tub.

Door switch
Prevents dishwasher from running when door is open.

Cycle extender switch
Usually operates a POTS AND PANS cycle, an extra feature on some models.

clog filters and small, hard objects such as olive pits can damage the plumbing. Read the dishwasher's Use and Care manual for directions about arranging dishes in the machine.

Water temperature is critical. For clean dishes, the water must be between 140°F and 160°F; a lower temperature won't dissolve grease or detergent. The water pressure to the dishwasher must be adequate, too. If pressure is too low, run the dishwasher only when no water is being used elsewhere in the house.

Your use of detergent can also affect how well your dishes come clean. Avoid using old dishwashing detergent; it can become ineffective in as little as two weeks after the foil seal is broken. If the water in your area is hard—high in mineral content—you may need to use more detergent. If you have soft water, or use a water softener, you may need to use less deter-

gent. Your machine may also have a rinse agent dispenser; be sure it is filled. A rinse agent makes water flow off dishes faster than normal, reducing water spotting.

The most difficult dishwasher repair involves servicing the pump and motor. An improperly installed pump seal, for example, can cause a leak into the motor, severely damaging the dishwasher. Therefore, only simple testing and removal of the pump-and-motor assembly has been shown *(page 84)*. If you find that the pump or motor is faulty, either remove the entire assembly and take it for repair, replace it with a new assembly, or call for professional service.

When repairing a dishwasher, always turn off the power at the house's main service panel or, for portable models, unplug the machine.

TROUBLESHOOTING GUIDE
continued ▶

SYMPTOM	POSSIBLE CAUSE	PROCEDURE
Dishes dirty or spotted after washing	Dishes loaded incorrectly	Rearrange dishes following manufacturer's loading instructions
	Water not hot enough	Test water temperature (p. 81) □○; if lower than 140°F, raise temperature at water heater
	Water pressure too low	Check water pressure (p. 82) □○; if low, avoid using house water supply while dishwasher is running
	Detergent ineffective	Make sure detergent is made for dishwashers; try different brands to find one effective for local water conditions
	Detergent dispenser faulty	Check for binding or broken parts (p. 78) □○
	Rinse agent dispenser empty	Refill dispenser
	Rinse agent leaking	Tighten loose fill cap; check rinse agent dispenser washers or gaskets; replace cracked tank (p. 78) □○
	Rinse agent dispenser faulty	Test bimetal terminals on rinse agent dispenser (p. 78) ◨●
	Spray arm stuck or clogged	Look for obstructions, such as measuring spoons, which fall under racks and block sprayer; check and clean spray arm (p. 80) □○
	Heating element faulty	Test heating element (p. 81) ◨●▲
	Selector switch or timer faulty	Test selector switch (p. 75) ◨●▲ and timer (p. 76) ◨●▲
	Pump clogged; impeller corroded, worn or chipped	Call for service
Dishwasher doesn't fill with water	Water supply line turned off or blocked	Check whether water is coming to faucet sharing same water line as dishwasher; if not, turn on water supply, clear sink drain clog, or call plumber
	Water inlet valve faulty or valve screen clogged	Test water inlet valve solenoid ◨●▲, and inspect filter screen (p. 83) ◨●
	Float or float switch jammed or faulty	Inspect float; test float switch (p. 82) ◨●▲
	Door switch faulty	Adjust door catch; test door switch (p. 77) ◨●▲
	With a low buzzing sound: Filter under spray arm clogged (models with filter)	Clean filter under faucet, using an old toothbrush □○
	Faucet coupler clogged (portable models)	If faucet coupler has screen, remove and clean or replace □○
Dishwasher drains during fill	Drain valve stuck open	Inspect drain valve; test-drain dishwasher with valve solenoid open and replace if necessary (p. 83) ◨●▲
Timer doesn't advance	Timer faulty	Test timer (p. 76) ◨●▲

DEGREE OF DIFFICULTY:	□ Easy ◨ Moderate ■ Complex	
ESTIMATED TIME:	○ Less than 1 hour ◑ 1 to 3 hours ● Over 3 hours	▲ Multitester required

TROUBLESHOOTING GUIDE (continued)

SYMPTOM	POSSIBLE CAUSE	PROCEDURE
Water doesn't shut off	Float or float switch jammed or faulty	Inspect float; test float switch (p. 82) ▣◗▲
	Faulty water inlet valve solenoid	Test water inlet valve solenoid (p. 83) ▣◗▲
	Clogged water inlet valve screen	Remove water inlet valve; clean or replace screen (p. 83) ▣◗
	Timer faulty	Test timer (p. 76) ▣◗▲
Motor doesn't run	No power to dishwasher	Check for blown fuse or tripped circuit breaker (p. 132) □○
	Door switch faulty	Adjust door catch; test door switch (p. 77) ▣◗▲
	Timer faulty	Test timer (p. 76) ▣◗▲
	Motor faulty	Test motor (p. 85) ▣●▲; remove pump-and-motor assembly for service, or call for service
Motor hums, but doesn't run	No power to dishwasher	Check for blown fuse or tripped circuit breaker (p.132) □○
	Door switch faulty	Adjust door catch; test door switch (p. 77) ▣◗▲
	Timer faulty	Test timer (p. 76) ▣◗▲
	Motor faulty	Test motor (p. 85) ▣●▲; remove pump-and-motor assembly for service, or call for service
	Impeller jammed (older models)	Call for service
Poor water drainage	Air gap clogged	Clean air gap (p. 79) □○
	Spray arm or filter screen clogged	Clean spray arm (p. 80) □○; if model has filter screen underneath spray arm, inspect and clean
	Drain hose blocked	Inspect drain hose (p. 79) ▣◗
	Drain valve solenoid faulty	Test drain valve solenoid (p. 83) ▣◗▲
	Timer faulty	Test timer (p. 76) ▣◗▲
	Pump impeller clogged or broken	Call for service
	Reversing motor doesn't reverse	Check motor for number of wires (p. 83) □○; only motors with four wires are reversible. Call for service
Dishwasher leaks around door	Use of wrong detergent	Use detergent recommended for dishwashers; do not prewash dishes with liquid detergent
	Dishes deflecting water through door vent	Reposition dishes in racks
	Door not closing tightly on gasket	Adjust door catch (p. 77) □○
	Door gasket hardened or damaged	Replace door gasket (p. 78) □○
Dishwasher leaks from bottom	Tub cracked	Seal crack with silicone rubber sealant or epoxy glue □○
	Water inlet valve connection loose	Tighten water inlet connection (p. 83) □○
	Hose split; hose clamp loose	Inspect hoses; adjust or replace clamps (pp. 79, 83) ▣◗
	Spray arm broken	Check spray arm (p. 80) □○
	Pump seal faulty	Call for service
Dishwasher doesn't turn off	Timer faulty	Test timer (p. 76) ▣◗▲
	Cycle extender switch faulty	Test cycle extender switch (p. 75) ▣◗▲
Dishwasher noisy	Dishes loaded improperly	Reposition dishes according to manufacturer's recommendations
	Water pressure too low	Check water pressure (p. 82) □○; if low, avoid using house water supply while dishwasher is running
	Water inlet valve screen clogged	Check water inlet valve screen (p. 83) ▣◗
	With a knocking sound during fill: Water inlet valve faulty	Test water inlet valve solenoid (p. 83) ▣◗▲
Door drops hard when opened	Door springs worn	Replace door springs (p. 77) □◗
Door is difficult to close	Door catch not closing properly	Adjust door catch (p. 77) □○
	Door catch bent or broken	Replace catch (p. 77) ▣◗

DEGREE OF DIFFICULTY: □ **Easy** ▣ **Moderate** ■ **Complex**
ESTIMATED TIME: ○ **Less than 1 hour** ◗ **1 to 3 hours** ● **Over 3 hours**

▲ **Multitester required**

ACCESS THROUGH THE CONTROL PANEL

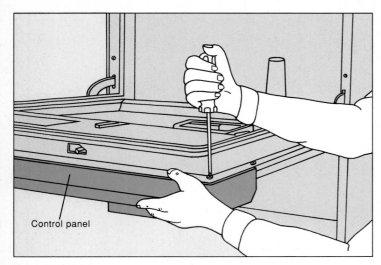

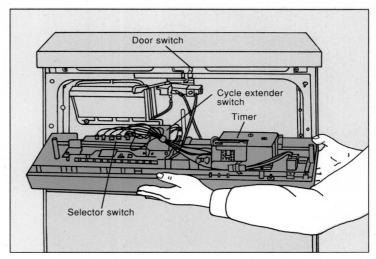

1 **Unscrewing the control panel.** Turn off power to the dishwasher. Remove the control panel retaining screws located, in most cases, inside the dishwasher door. Hold the control panel to keep it from falling.

2 **Freeing the control panel.** Close the dishwasher door. Supporting the panel to avoid damaging wires, lower it away from the door, as shown. (On some machines, you may first have to remove the door handle or door panel.) You now have access to the selector switch, cycle extender switch, door switch and timer.

ACCESS THROUGH THE DOOR PANEL

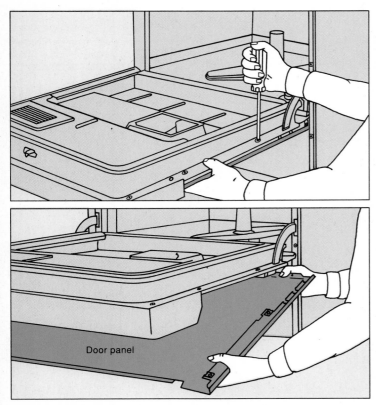

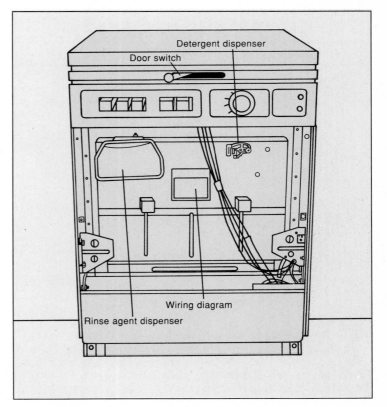

1 **Removing the door panel.** Turn off power to the dishwasher. Open the door and, keeping one hand underneath it to support the panel, remove the retaining screws *(above, top)* and lower the panel away *(above, bottom)*. On some dishwashers the screws are hidden under metal strips; lift the strip to reach the screws.

2 **Getting to the door panel parts.** With the control panel and door panel removed, you now have access to the door switch, detergent and rinse agent dispensers, and the wiring diagram.

ACCESS THROUGH THE LOWER PANEL

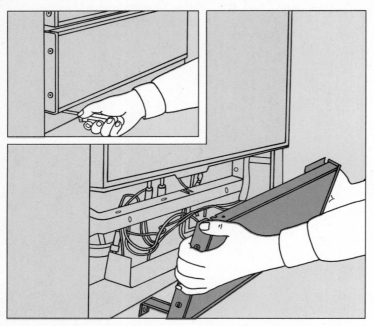

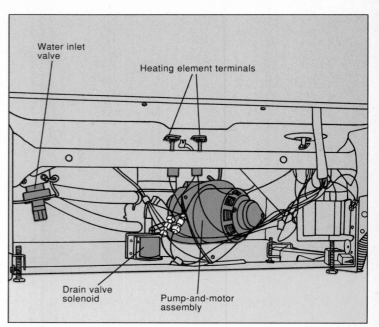

Water inlet valve

Heating element terminals

Drain valve solenoid

Pump-and-motor assembly

1 **Removing the lower panel.** Turn off power to the dishwasher. Depending on the model, remove any retaining screws *(inset)* and, to free the panel, pull it down or lift it off hooks *(above).*

2 **Getting to parts under the tub.** With the lower panel removed, you have access to the water inlet valve, heating element terminals, pump-and-motor assembly and drain valve solenoid. Some of these parts are hard to reach, so for complicated repairs on a built-in model, pull the machine free from the counter and tilt it on its back *(below).*

FREEING AND TILTING THE MACHINE

Water inlet valve

Leveling foot

1 **Disconnecting the power and water supply.** Turn off the water and power supply at their sources. Open the door and remove the screws that secure the top of the dishwasher to the kitchen counter. Remove the lower panel *(step 1, above),* then turn the threaded leveling feet to lower the dishwasher slightly. Keeping a shallow pan handy to catch dripping water, use a wrench to disconnect the water supply line from the water inlet valve *(above).* Disconnect the drain hose from the sink drain. Unscrew the cover of the power cord junction box, near floor level, and disconnect the power cord.

2 **Tilting the dishwasher.** Slide a blanket beneath the machine to protect the floor from scratches, then rock the appliance clear of the cabinet. Make sure you have a steady grip on the machine, and with your legs bent, gently lower it backward *(above).*

SERVICING THE SELECTOR SWITCH

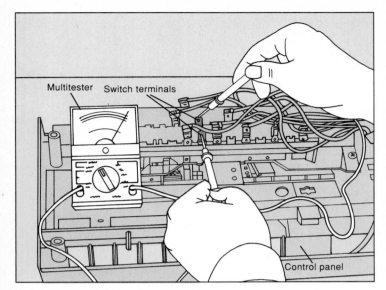

Multitester   Switch terminals

Control panel

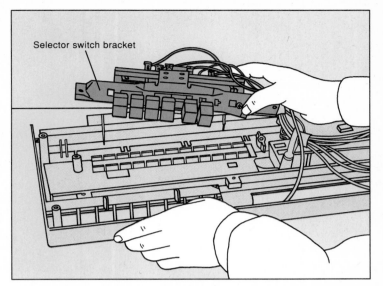

Selector switch bracket

1 **Testing the selector switch.** After turning off the power or unplugging the power cord, remove the control panel *(page 73)* and inspect the selector switch. Repair any loose or broken wires *(page 136)*. Refer to the wiring diagram to test each pair of switch terminals for continuity, using a continuity tester or a multitester set at RX1 *(above)*. To make sure you reconnect the wires properly, disconnect only two at a time, reconnecting them before moving on to the next set. If any pair of terminals fails to show continuity with the appropriate switch button pressed, remove and replace the switch.

2 **Replacing the selector switch.** Detach the selector switch bracket from the control panel by removing the mounting screws and lifting the bracket from the control panel *(above)*. Pull the buttons off, keeping them in order for reinstallation. Unscrew the faulty switch from its bracket, leaving the wires attached. Install a new switch and replace the pushbuttons. Finally, transfer the wires one by one from the old switch to the same terminals on the new switch. Replace the bracket in the control panel.

SERVICING THE CYCLE EXTENDER SWITCH

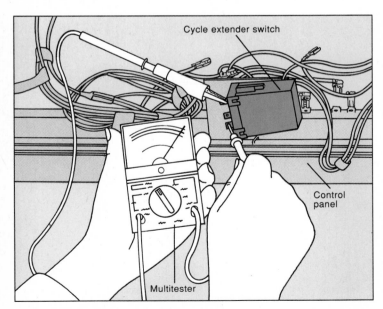

Cycle extender switch

Control panel

Multitester

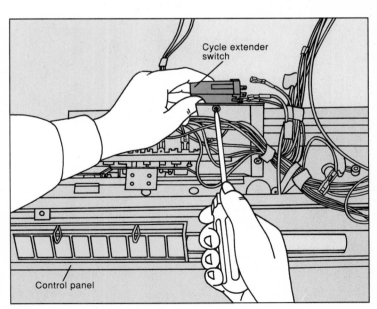

Cycle extender switch

Control panel

1 **Testing the cycle extender switch.** Turn off power to the dishwasher, remove the control panel *(page 73)* and inspect the switch. Repair any loose or broken wires *(page 136)*. Disconnect the wires from the switch terminals and label their positions. With a multitester set at RX1, test terminals H2 and L2 for continuity, as shown. Then set the multitester at RX100 and test terminals H2 and H1; they should show resistance. If the switch is faulty, replace it.

2 **Replacing the cycle extender switch.** Depending on your model, unscrew *(above)* or unclip the switch from the control panel. Install a new switch, reconnecting the wires one by one to the new terminals, and reinstall the control panel.

SERVICING THE TIMER

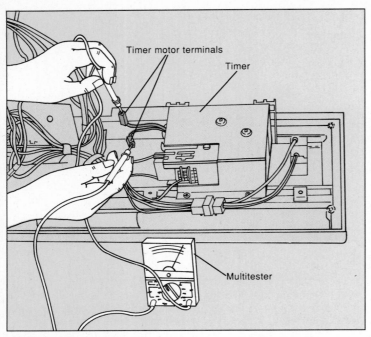

Timer motor terminals

Timer

Multitester

1 **Testing the timer motor.** After turning off the power or unplugging the power cord, remove the control panel *(page 73)*. Disconnect the timer motor wires and, using a multitester set at RX100, test the timer motor terminals *(above)*; they should show partial resistance. If the motor is faulty, replace the entire timer *(step 4)*. If not, next check the timer plug and terminals.

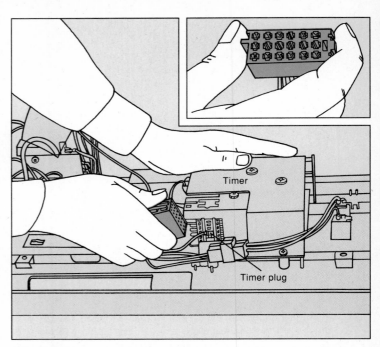

Timer

Timer plug

2 **Disconnecting the timer plug.** Keeping a solid grip on the timer plug, wiggle it away from the timer *(above)*. Once it is removed, inspect the plug *(inset)* for loose contacts. If any contacts protrude from the plug, push them back into place.

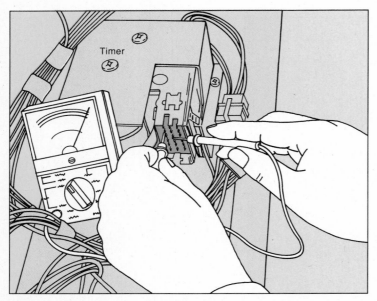

Timer

3 **Testing the timer terminals.** For this step, refer to the dishwasher's wiring diagram and timer chart, found inside the door panel *(page 73)* or available from the manufacturer. Use the wiring diagram to identify the pairs of terminals that control the faulty cycle. Disconnect the wires from those terminals and test each pair for continuity with a multitester set at RX1 *(above)*. Turn the timer dial slowly through the full cycle. The timer chart will show at which points in the cycle there should be continuity. To make sure the wires are properly reconnected, replace each pair before disconnecting another pair. If any pair of terminals does not show continuity, replace the timer.

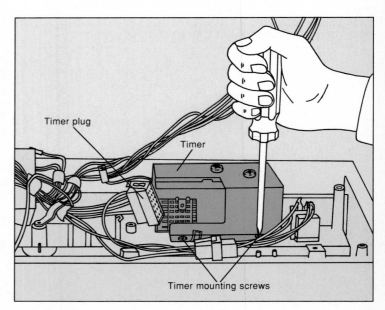

Timer plug

Timer

Timer mounting screws

4 **Replacing the timer.** Pull off the timer dial by hand, then remove the timer mounting screws *(above)*. Disconnect all other wires, including the motor wires, and mark them with masking tape for reassembly. Install a new timer, reconnect the wires and reinstall the timer plug. Replace the dial and the control panel.

SERVICING THE DOOR SWITCH

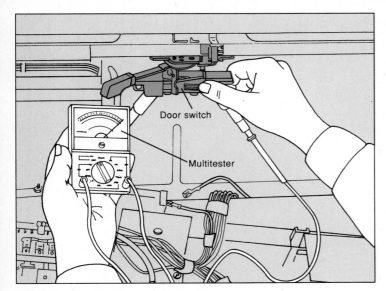

Door switch

Multitester

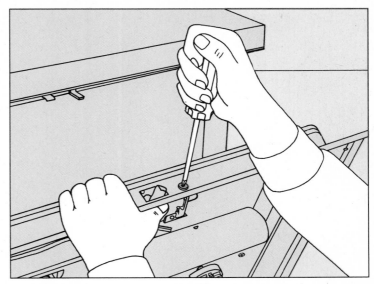

1 **Testing the door switch.** Before testing the door switch, check to see if the door catch closes securely. On many models, the catch can be tightened or repositioned by loosening the retaining screws, sliding the catch in or out, then retightening the screws. Otherwise, shut off the power and remove the control panel *(page 73)*. With the door closed and locked, disconnect the wires from the door switch terminals and test for continuity using a multitester set at RX1 *(above)*. If there is no continuity, replace the switch.

2 **Replacing the door switch.** With the wires disconnected, remove the door switch retaining screws *(above)*, install a new switch and reconnect the wires.

REPLACING THE DOOR SPRINGS

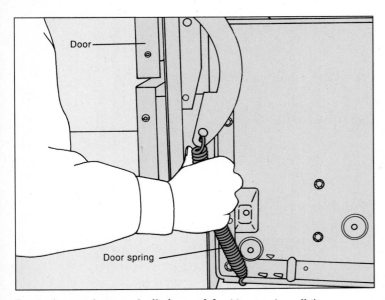

Door

Door spring

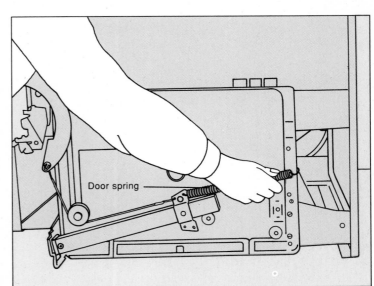

Door spring

Replacing springs on built-in models. After turning off the power supply, locate the door springs. They are usually on the sides of the machine, under the front corners of the tub, so pull the dishwasher out from the counter *(page 74)*. If the springs are weak or broken, remove the ends from their holes by hand and replace them as shown. For proper tension, always replace door springs in pairs, even if only one is broken. If your dishwasher has a series of holes next to the pair on which the springs are hooked, you can adjust the tension by hooking each spring into a different hole.

Replacing springs on portable models. After unplugging the machine, remove the side panels by unscrewing the retaining screws that secure them to the frame. The springs are attached to an anti-tip mechanism that prevents the machine from falling forward when the door is opened and the racks pulled out. If the springs are weak or broken, replace them by hooking new springs into position *(above)*. For proper tension, always replace springs in pairs, even if only one is broken.

REPLACING THE DOOR GASKET

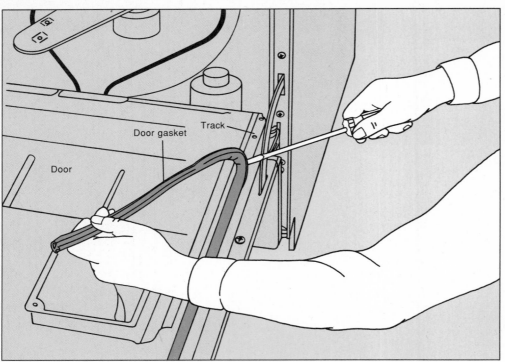

Inspecting and replacing the door gasket. Turn off power to the dishwasher. Open the dishwasher door, remove the dish racks and inspect the door gasket. If it is cracked or otherwise damaged, replace it with an identical gasket. Some may also have a tub gasket for extra protection; check this, too.

On some dishwashers, the gasket has clips or tabs that can be pried out with a screwdriver *(left)*. Other gaskets are secured with retaining screws. If the replacement gasket is kinked, soak it in warm water for a few minutes before installation. If the gasket slides into a track, lubricate the gasket with soapy water or silicone to make installation easier. (Do not use oil or grease.)

Press the center of the gasket into the top center of the door. Continue around the door, pressing into place several inches at a time. Secure the ends with the original clips or brackets.

If the gasket is good, check to see if the door catch closes securely. On many models, the catch can be tightened or repositioned by loosening the retaining screws, sliding the catch in or out, then retightening the screws.

CHECKING THE DETERGENT AND RINSE AGENT DISPENSERS

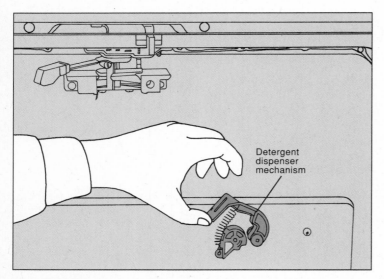

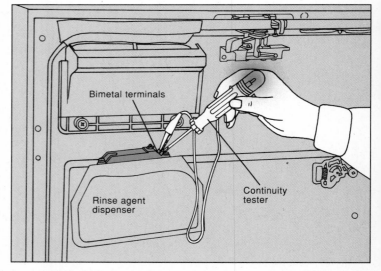

Inspecting the detergent dispenser. Although their design may differ, all detergent dispensers operate in much the same way. The dispenser is opened mechanically by the timer at a predetermined point during the cycle. Breakdowns, usually due to a damaged part, are infrequent. First turn off the power or unplug the dishwasher, then check the detergent cup inside the door for caked detergent; clean it if necessary. Also check the O-ring or gasket, if any, inside the cup's cover, and replace it if damaged. If your model has a movable cup, open and close it by hand to see if it is stuck. Next remove the door panel *(page 73)* and check the spring-and-lever mechanism for stuck or broken parts *(above)*. Replace any part that is damaged.

Inspecting the rinse agent dispenser. If your dishwasher has a rinse agent dispenser, the rinse agent is released either by an arm on the detergent dispenser or by the timer. First turn off the power to the dishwasher and remove the door panel *(page 73)*. Make sure the fill cap is tight and inspect the rinse agent dispenser for signs of damage. If it is split or cracked, replace it. If your model resembles the one shown above, remove the wires to the bimetal terminals and test the terminals for continuity, as shown. If there is no continuity, replace the bimetal assembly. If the dispenser leaks, check the washers or gaskets and replace them if cracked or brittle.

UNCLOGGING THE AIR GAP (Built-in models)

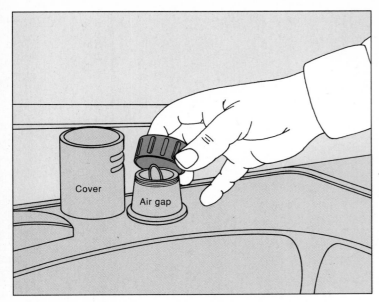

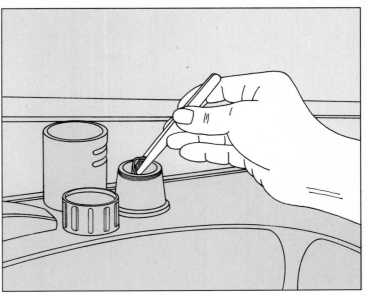

1 **Access to the air gap.** This plumbing part, required by many local building codes, prevents water in the kitchen sink or garbage disposer from backing up into the dishwasher. Pull the cover off the air gap, located next to the sink faucet, and unscrew the cap under the cover, as shown. (On portable models, the faucet aerator serves as an air gap, since the sink faucet is higher than the dishwasher.)

2 **Cleaning the air gap.** With tweezers, remove any debris such as glass, bones or toothpicks from the small tube in the center of the air gap *(above)*. Rinse the cover and cap and reassemble.

INSPECTING AND REPLACING THE DRAIN HOSE

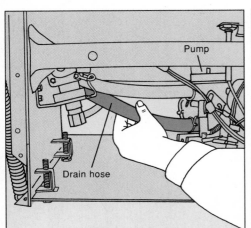

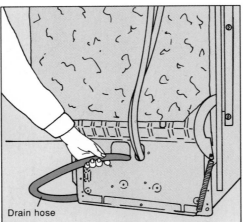

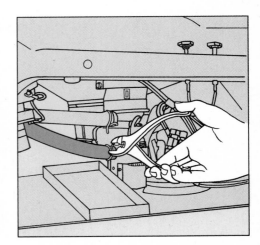

1 **Inspecting the drain hose.** Turn off the power and water supply to the dishwasher. Remove the lower panel *(page 74)* and inspect the drain hose for kinks, cracks or splits. Begin at the pump, straightening any kinks by hand *(above, left)*. If the kink remains, replace the hose. Also check that the hose clamp is secure; if not, reposition it with pliers.

If you don't find the problem under the dishwasher, follow the hose to the kitchen sink drain or garbage disposer (built-in models, *above, right*) or the faucet coupler (portable models). This means gaining access to the side of the dishwasher, so pull a built-in model away from the counter *(page 74)* or remove the side and top panels of a portable model *(page 77)*. If the hose is collapsed, kinked, or cracked, replace it.

2 **Replacing the drain hose.** With a shallow pan handy to catch dripping water, disconnect the hose from the pump by squeezing the spring clamp with pliers *(above)*. Then disconnect the hose under the kitchen sink drain (built-in models) or faucet coupler (portable models). Position the new hose and reconnect it at both ends with new clamps.

SERVICING THE SPRAY ARM

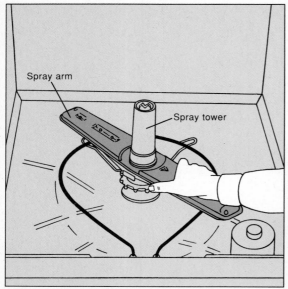

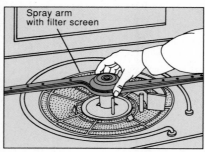

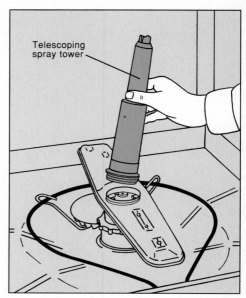

1 **Checking for obstructions.** Turn off power to the dishwasher and slide out the lower dish rack. Rotate the spray arm to see if it moves freely *(above)*; the ends of the arm should also move up and down slightly. If bent or damaged, replace the spray arm.

2 **Removing the spray tower.** If your model has a spray tower in addition to the spray arm, pull the telescoping parts to see if they move freely. Unscrew the tower by hand and remove it, as shown. Check for clogs and clean if necessary.

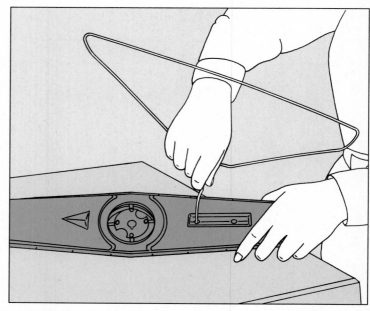

3 **Removing and replacing the spray arm.** Remove the spray arm by unscrewing it, as shown, or by removing a plastic cap. Keeping them in order, lift out the spray arm and its washers, gaskets, or bearings. Clean or, if it is bent or damaged, replace the spray arm. If your model has a strainer below the spray arm, remove it and clean with an old toothbrush under running water.

4 **Cleaning the spray arm.** Check the spray arm for foreign objects. Using a stiff wire bent at a right angle at the tip, unclog the holes *(above)*. If your model has a second spray arm mounted above the first, clean it, too. Then rinse the spray arm, tower and strainer under running water and reassemble the parts.

MEASURING THE WATER TEMPERATURE

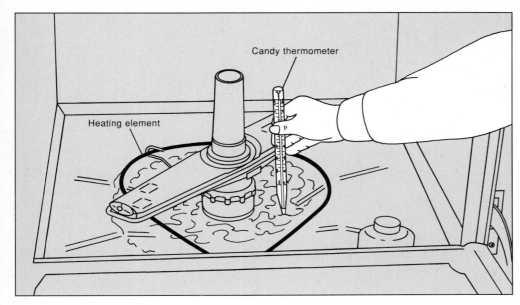

Candy thermometer

Heating element

Measuring the water temperature. Turn on the dishwasher, then interrupt it during the first wash cycle by opening the door. Steam will pour out, but the water will remain in the bottom of the tub. Place a candy or meat thermometer in the water, as shown. It should read at least 140°F. If not, raise the temperature slightly at your water heater thermostat and test again after one hour.

If your water heater is already set to a higher temperature than is being delivered to the dishwasher, keep in mind that a loss of one degree per foot of pipe is normal. If the water temperature in the dishwasher continues to stay below 140°F, check the heating element *(below)*.

SERVICING THE HEATING ELEMENT

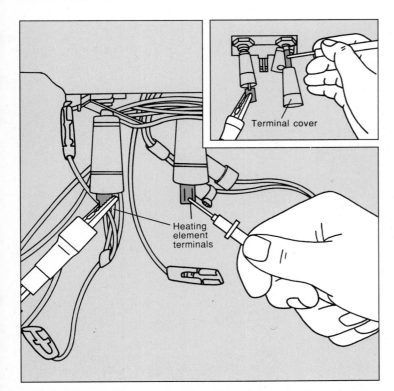

Terminal cover

Heating element terminals

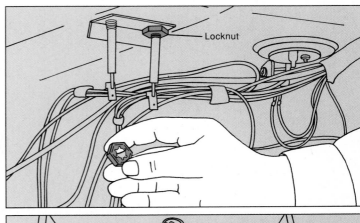

Locknut

1 **Testing the heating element.** Turn off power to the dishwasher and remove the lower access panel *(page 74)*. Disconnect the heating element terminal wires, located under the tub, and test the terminals for continuity, using a multitester set at RX1 *(above)*. The tester should indicate partial resistance. If not, replace the element.

If the test shows continuity, check the element for a ground. Pull one of the rubber covers to the end of the terminal. Touch one probe of the tester to the metal sheath and the other to a terminal *(inset)*. A continuity reading means the heating element should be replaced.

2 **Replacing the heating element.** Slide off the rubber terminal covers and remove the locknuts that hold the element in place under the tub *(top)*. Then, from inside the tub, remove the element from its bracket *(bottom)*. Lift the element from the tub and install the exact replacement for your make and model. Reconnect the wires to the terminals.

CHECKING THE WATER PRESSURE

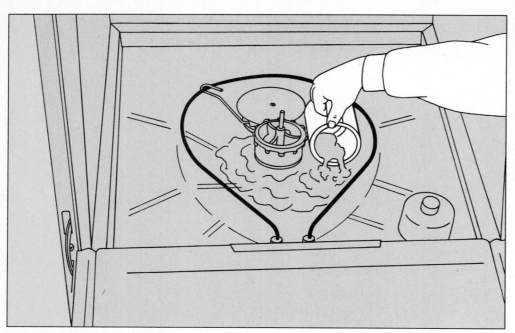

Checking the water pressure. Turn on the dishwasher and let it run until the dial reaches the first wash cycle. Then stop the machine by opening the door. Steam will pour out, but the water will remain in the bottom of the tub. Let the water cool, then bail it out *(left)* into a gallon container—emptying it into the sink as it becomes full. Remove the spray arm *(page 80)* if it is in your way. If there is less than 2 1/2 gallons (2 imperial gallons) of water in your dishwasher, the water pressure is probably too low. To remedy the problem, avoid using the house water supply while the dishwasher is in use.

REPAIRING THE FLOAT SWITCH

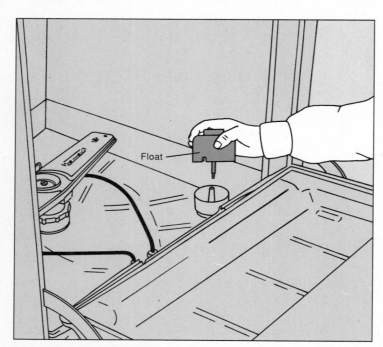

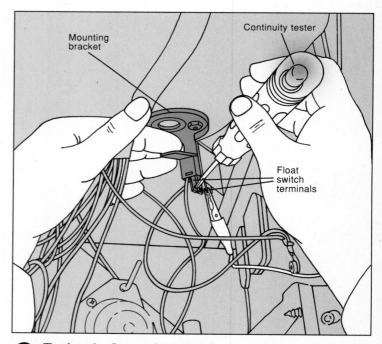

1 **Inspecting the float.** Turn off power to the dishwasher. Open the dishwasher door, remove the lower dish rack and jiggle the float up and down to check that it moves freely. Then pull it out and look for obstructions *(above)*. On some models, the float is held in place by a clip under the tub. Remove the lower panel *(page 74)* and remove the clip. On other models, the float is hidden by a cover that first must be removed. Replace the float with a new one if it is damaged. If the float moves freely but the machine either doesn't fill or overflows, test the float switch.

2 **Testing the float switch.** Remove the lower panel *(page 74)* and detach the wires from the float switch terminals. Put the switch in the ON position by pulling down on the lever, and test the terminals for continuity. If there is no continuity, remove the switch by unscrewing it from its mounting bracket and replace it.

REPAIRING THE WATER INLET VALVE

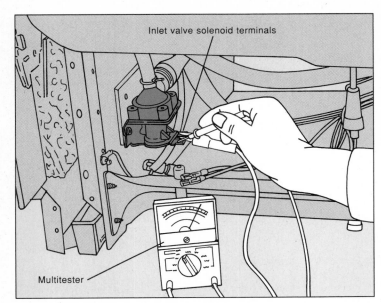

Inlet valve solenoid terminals

Multitester

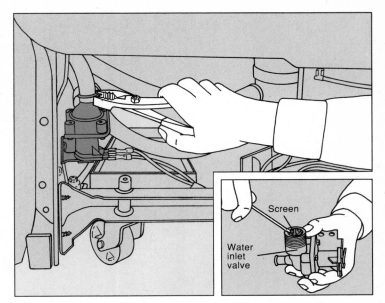

Screen

Water inlet valve

1 **Inspecting the water inlet valve solenoid.** Turn off the power and water supply to the dishwasher. Remove the lower panel *(page 74)*; you may also pull out the machine and tip it back for easier access. (On some portable models, remove the top of the machine to locate the water inlet valve.)

If your machine leaks, make sure the incoming water line and the hose that connects the inlet valve to the tub are securely fastened. Tighten, if necessary, using an adjustable wrench or hose-clamp pliers.

If you suspect the valve is faulty, remove the wires from the inlet valve solenoid terminals and test for continuity with a multitester set at RX1, as shown. If there is no continuity, remove and service the valve.

2 **Servicing the water inlet valve.** With a shallow pan handy to catch dripping water, use hose-clamp pliers to remove the hose that connects the inlet valve to the tub, as shown. Then use a wrench to disconnect the incoming water line. To free the water inlet valve, remove any screws that secure the valve bracket to the tub. If the valve is cracked or otherwise damaged, replace it.

Using a small screwdriver, pry out the water inlet valve screen *(inset)*. If it is a plastic screen, rinse it and clean it with an old toothbrush, then replace it in the valve. If the screen is metal, it won't retain its shape; replace it with a new one. Reinstall the valve, water line and hose and check all connections.

REPAIRING THE DRAIN VALVE

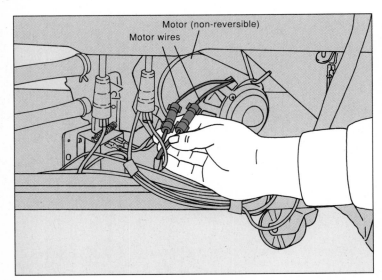

Motor (non-reversible)
Motor wires

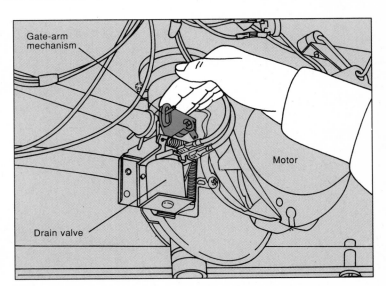

Gate-arm mechanism

Motor

Drain valve

1 **Locating the drain valve.** Drain valves are used in dishwashers with non-reversible motors. Turn off power to the dishwasher and remove the lower panel *(page 74)*. Count the number of wires attached to the motor. A motor with two or three wires is non-reversible, as above; one with four wires is reversible.

2 **Checking the gate-arm mechanism.** Tilt the machine on its back *(page 74)* to make it easier to reach the drain valve. If your drain valve has a gate-arm mechanism, move it by hand *(above)*; it should move freely up and down. There should be two springs; replace any that are broken or missing.

REPAIRING THE DRAIN VALVE (continued)

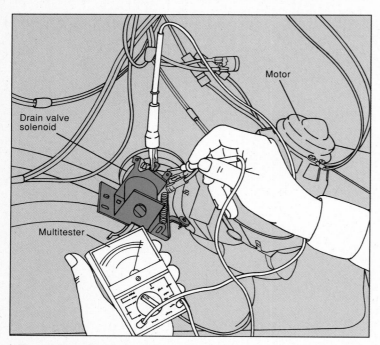

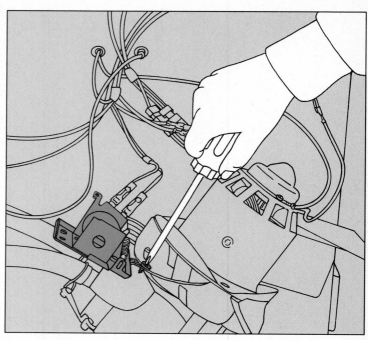

3 **Testing the drain valve solenoid.** Disconnect the wires from the drain valve solenoid terminals and, using a multitester set at RX1, test it for continuity *(above)*; the solenoid should show partial resistance. If there is no continuity, replace the solenoid.

4 **Replacing the drain valve solenoid.** Noting carefully how they are attached, remove the mounting screws *(above)* and detach the solenoid springs and wires. Install a new solenoid and reattach the wires.

SERVICING THE PUMP-AND-MOTOR ASSEMBLY

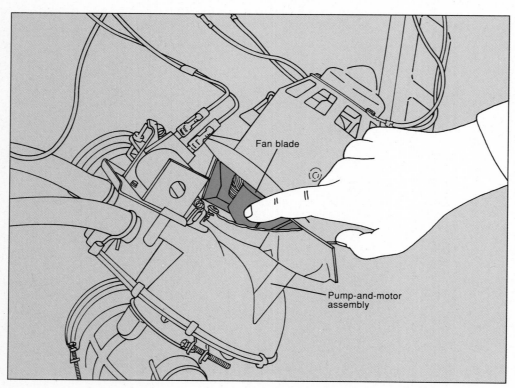

1 **Inspecting the motor fan blades.** Turn off power to the dishwasher and remove the lower access panel *(page 74)*. If it is a built-in model, pull the dishwasher away from the cabinet and tilt it on its back *(page 74)* for access to the pump-and-motor assembly. If the motor fan blades are visible on your model, turn them by hand *(left)*; they should move freely. If not, the pump seal may be stuck or an object may be lodged in the assembly. Check for obstructions; if none are visible, remove the assembly or call for professional service.

SERVICING THE PUMP-AND-MOTOR ASSEMBLY (continued)

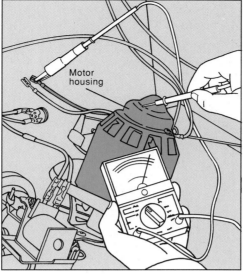

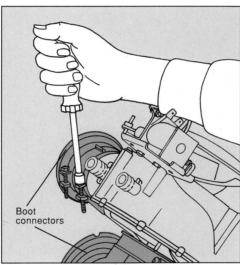

2 **Testing the motor.** Snap off any protective brackets and disconnect the motor wires from the motor terminals. Test for continuity *(far left)*; with the multitester set at RX1, the motor should show partial resistance. If there is no continuity, replace the pump-and-motor assembly or call for service.

If the motor shows continuity, test next for a ground. Place one probe of the multitester on the bare metal motor housing, the other probe on each terminal in turn *(near left)*. A continuity reading indicates a ground; replace the pump-and-motor assembly or call for service.

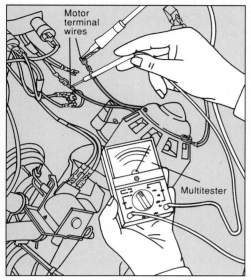

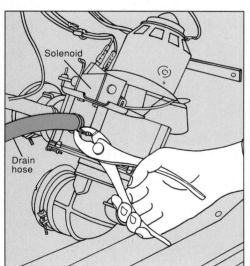

3 **Disconnecting the pump-and-motor assembly.** With a shallow pan handy to catch dripping water, use hose-clamp pliers to disconnect the hoses from the pump *(far left)*. Mark the hose positions for reconnection. Detach the drain valve solenoid wires, then unscrew the clamp closest to the pump on each boot connector *(near left)*.

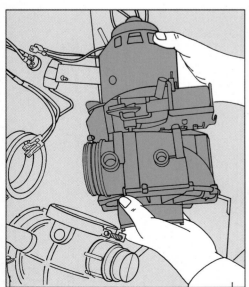

4 **Replacing the pump-and-motor assembly.** Remove the ground wire from the assembly and, on models with access under the tub, pull the assembly free from its hanger *(far left)*. On other models, remove the assembly from inside the tub *(near left)*. You now have access to the pump. Although there may be pump repair kits available for your model, attempt such a repair only if you are a highly experienced do-it-yourselfer. Otherwise, replace the entire assembly or take it to a service center for repair.

GARBAGE DISPOSERS

A garbage disposer is built to do a dirty job, efficiently grinding and shredding such tough food waste as melon rinds, chicken bones, fruit pits and vegetable parings. Used properly, the disposer is a rugged appliance that should remain trouble-free for years. The most common disposer problem is a simple jam and even more complex fixes, such as replacing a dull shredder ring, are easily handled by the do-it-yourselfer.

Garbage is fed into the disposer through the sink drain and collects in the hopper (waste may also come through a hose from the dishwasher drain). Batch-feed disposers are activated by inserting and turning the stopper in the sink drain opening. Continuous-feed models are controlled by a wall switch. On both models, the force of the spinning flywheel, which is attached to the motor shaft, propels waste against the shredder ring, catching it with the impellers and grinding it into small pieces. Running water flushes the ground-up waste through openings in the flywheel and down the drain.

To avoid problems with the disposer, read its Use and Care manual and follow the manufacturer's instructions for the proper amount and kinds of food to be placed in it. To speed up disposal, cut or break up large bones, corncobs, and fibrous material such as vegetable stalks before putting them into the disposer. Drop in garbage loosely; packing it tightly could jam the disposer. Never put in hard-to-grind waste such as seafood

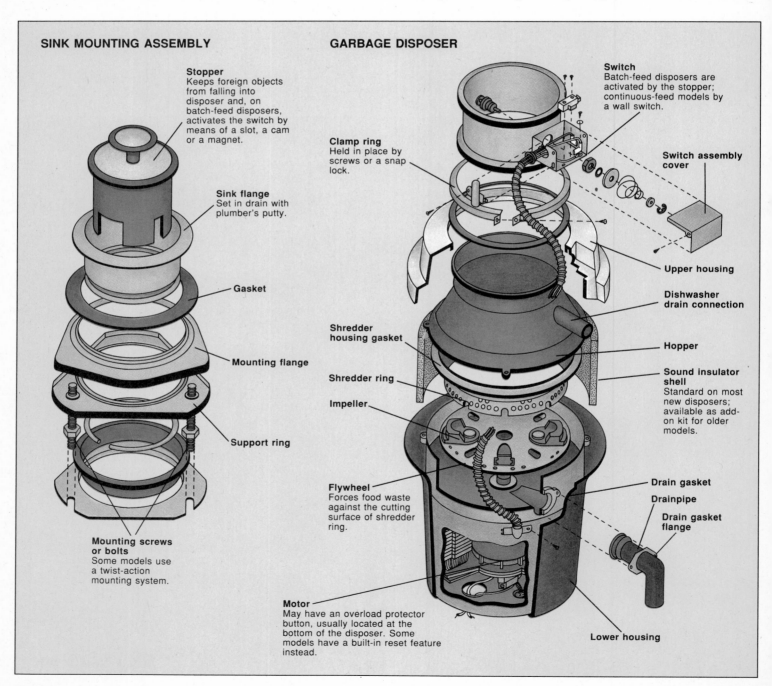

SINK MOUNTING ASSEMBLY

Stopper
Keeps foreign objects from falling into disposer and, on batch-feed disposers, activates the switch by means of a slot, a cam or a magnet.

Sink flange
Set in drain with plumber's putty.

Gasket

Mounting flange

Support ring

Mounting screws or bolts
Some models use a twist-action mounting system.

GARBAGE DISPOSER

Switch
Batch-feed disposers are activated by the stopper; continuous-feed models by a wall switch.

Switch assembly cover

Clamp ring
Held in place by screws or a snap lock.

Upper housing

Dishwasher drain connection

Shredder housing gasket

Hopper

Shredder ring

Sound insulator shell
Standard on most new disposers; available as add-on kit for older models.

Impeller

Flywheel
Forces food waste against the cutting surface of shredder ring.

Drain gasket

Drainpipe

Drain gasket flange

Motor
May have an overload protector button, usually located at the bottom of the disposer. Some models have a built-in reset feature instead.

Lower housing

shells, artichoke leaves or corn husks unless the manufacturer says that the disposer will accept them. These can make the motor overheat, tripping the overload protector. Never put in non-food waste such as metal, cloth, plastic, rubber or ceramics, which can seriously damage the disposer or plumbing.

Always use a strong flow of cold water when running the disposer. This assists shredding and congeals grease so that it breaks up and floats away through the drain. After turning off the disposer, let the water run for a minute or so to flush the drain line.

Disposers require little maintenance. Do not use chemical drain cleaners to clear clogs; these can damage plastic and rubber parts. To keep the drain clean and odor-free, do not leave garbage in the disposer for more than a day. When necessary, deodorize the disposer by using it to grind an orange or lemon rind. When the disposer is not in use, leave the drain cover in place to prevent objects such as silverware or bottle caps from accidentally falling in and causing a jam.

Most disposer repairs require only simple household tools, but be sure to take safety precautions when working with the appliance. Never put your hand into the disposer, even when the unit is off; doing so could activate the switch. Do not attempt any repair, even freeing a simple jam, without first turning off power at the main service panel *(page 132)*.

TROUBLESHOOTING GUIDE

SYMPTOM	POSSIBLE CAUSE	PROCEDURE
Disposer doesn't work at all	No power to disposer	Check for blown fuse or tripped circuit breaker *(p. 132)* □○; for plug-in disposers, check for power at outlet by plugging in lamp
	Disposer jammed, tripping overload protector	Free flywheel, then push overload protector button *(p. 89)* □○
	Wall switch faulty (continuous-feed disposers)	Test wall switch *(p. 89)* ▄●
	Switch faulty (batch-feed disposers)	Test switch *(p. 90)* ▄●
	Motor faulty	Call for service
Disposer buzzes, but doesn't work	Flywheel jammed	Free flywheel *(p. 89)* □○
	Motor faulty	Call for service
Disposer drains poorly	Water flow insufficient	Open cold water faucet fully when operating disposer
	Drain line clogged	Disconnect and dismount the disposer *(p. 88)* and use a plumber's auger to clear the drain ▄●, or call a plumber; do not use chemical drain openers
	Shredder ring dull or flywheel broken	Inspect shredder ring and flywheel *(p. 90)* ▄●
Disposer will not stop	Wall switch faulty (continuous-feed disposers)	Test wall switch *(p. 89)* ▄●
	Switch faulty (batch-feed disposers)	Test switch *(p. 90)* ▄●
Disposer grinds too slowly	Water flow insufficient	Open cold water faucet fully when operating disposer
	Improper waste in disposer	Disconnect power to disposer; remove any waste not recommended for disposal by manufacturer □○
	Shredder ring dull or flywheel broken	Inspect shredder ring and flywheel *(p. 90)* ▄●
Disposer starts, but stops when stopper is released	Switch faulty or cam damaged (batch-feed disposers)	Test switch; if OK, inspect cam *(p. 90)* ▄●
Disposer leaks	Drain gasket flange loose	Tighten screws on drain gasket flange □○
	Poor seal at sink connection	Carefully tighten support ring screws or bolts (overtightening can crack porcelain sink); apply plumber's putty if necessary □○
	Poor seal between hopper and lower housing	Replace shredder housing gasket *(p. 91)* ▄●
Disposer vibrates or is unusually noisy	Silverware or other object in disposer	Disconnect power to disposer; remove object from disposer with pliers or kitchen tongs; push overload protector button *(p. 89)* □○
	Mounting system loose	Tighten loose mounting screws or bolts □○
	Flywheel damaged	Inspect flywheel *(p. 90)* ▄●
	Motor faulty	Call for service

DEGREE OF DIFFICULTY: □ **Easy** ▄ **Moderate** ■ **Complex**
ESTIMATED TIME: ○ **Less than 1 hour** ● **1 to 3 hours** ● **Over 3 hours**

DISCONNECTING AND DISMOUNTING THE DISPOSER

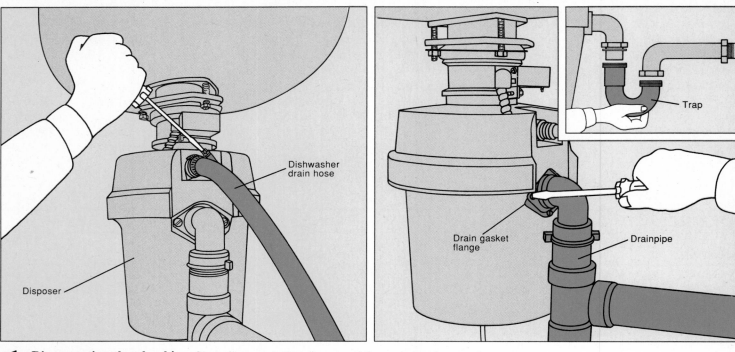

Dishwasher drain hose

Disposer

Trap

Drain gasket flange

Drainpipe

1 **Disconnecting the plumbing.** Shut off power to the disposer at the main service panel. Keeping a container handy to catch dripping water, separate the dishwasher drain hose, if any, from the disposer. On some models, the hose is detached using a screwdriver *(above, left);* on others use hose-clamp pliers. Next unscrew the drain gasket flange and pull out the drainpipe *(above, right)*. On others, loosen the slip nuts and drop the sink trap *(inset)*.

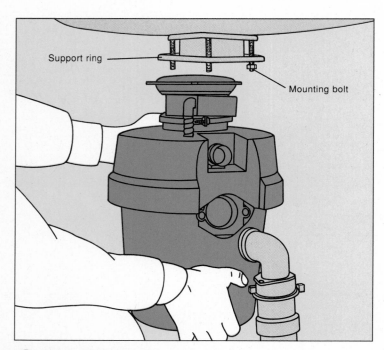

Support ring

Mounting bolt

2 **Dismounting the disposer.** Loosen the mounting screws or bolts at the top of the disposer. Then, keeping a hand under the disposer so that it won't fall, twist the unit free. Lower the disposer from the support ring, as shown. If your disposer is fastened with a twist-lock action, rotate it, to the left in most cases, until it slips free.

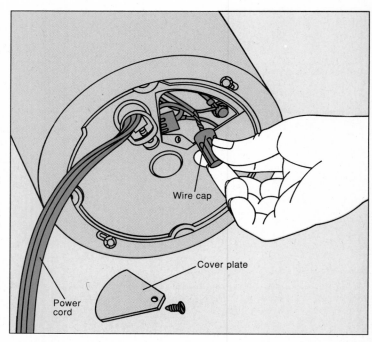

Wire cap

Cover plate

Power cord

3 **Disconnecting the power supply.** Unscrew the electrical cover plate on the bottom of the disposer and disconnect the green ground wire. Disconnect the white and black wires by unscrewing the two wire caps by hand, as shown, and pull the power cord free from the disposer.

FREEING A JAMMED DISPOSER

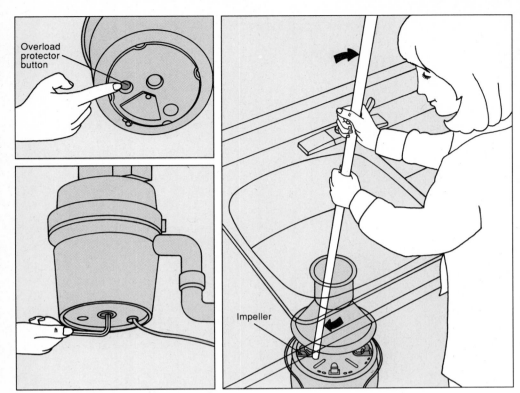

Overload protector button

Impeller

Freeing the flywheel. Disconnect power to the disposer. If the jam was caused by an object such as a piece of silverware, remove it from the disposer and wait 15 minutes to allow the motor to cool. Reconnect the power, push the red overload protector button on the bottom of the machine *(far left, top)* and try operating the disposer again.

If the disposer is still jammed, disconnect the power and wedge the end of a broom handle against one of the impellers on the flywheel *(near left)*. Using the handle as a lever, force the wheel back and forth until it moves freely. Wait for the motor to cool. Reconnect the power, press the overload protector button and try operating the disposer again. If your model has a manual reversing switch (usually located on the lower housing), turn the disposer off, then turn the reversing switch to help free the jam.

If your disposer came with a large hex wrench, insert one end of the wrench into the hexagonal hole at the bottom of the motor housing *(far left, bottom)*. Turn the wrench back and forth to rotate the motor shaft until the flywheel is free. Wait for the motor to cool. Push the overload protector button and try operating the disposer again.

TESTING AND REPLACING THE SWITCH (Continuous-feed models)

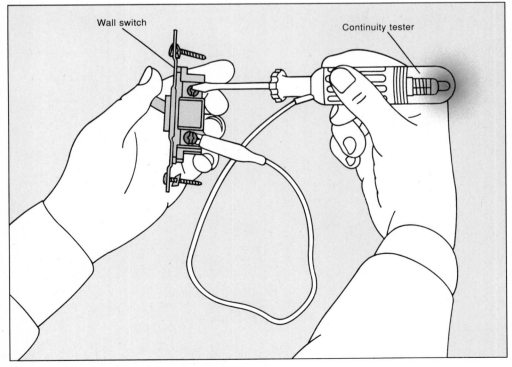

Wall switch

Continuity tester

Testing and replacing the wall switch. Turn off power to the switch at the main service panel. First check the switch toggle; if it is loose or won't stay in the ON position, the switch should be replaced. Next test the switch itself. Make sure the power supply is turned off, remove the cover plate and pull the switch from its box in the wall. Disconnect the terminal leads and, with the switch in the ON position, test for continuity, as shown. If the continuity tester's light does not glow, replace the switch. Reconnect the wires to a new switch, replace the switch in its box in the wall, and screw on the cover plate.

TESTING AND REPLACING THE SWITCH (Batch-feed models)

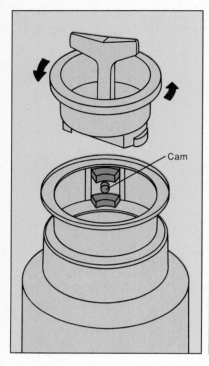

Cam

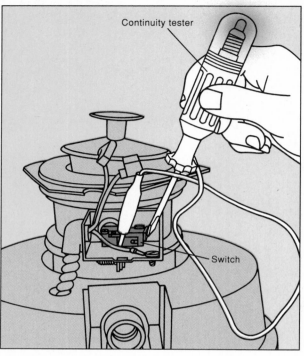

Continuity tester

Switch

Testing and replacing the stopper-activated switch. The switch on a batch-feed disposer is located on the disposer itself, near the neck. If your model uses a stopper with either a cam slot or a magnet to start the disposer, it probably has a switch like the one shown at near left. Although you can test the switch while the disposer is still mounted under the sink, it's easier first to disconnect and dismount the disposer *(page 88)*. Turn off the power supply, then unscrew and remove the switch assembly cover to reveal the switch. Disconnect the terminal leads and, with the switch in the ON position (stopper in), test for continuity *(near left)*. If the light does not glow, replace the switch. Remove the screws holding the switch in place, install a new switch, reconnect the terminal leads and replace the switch assembly cover.

If the switch tests OK, check to see if the disposer has a cam located on the neck *(far left)*. On older disposers, this cam activates the switch when the stopper is inserted and rotated in the neck of the machine. After some time the cam may wear. Inspect it for damage and, if necessary, unscrew and replace it.

REPLACING THE SHREDDER RING AND FLYWHEEL

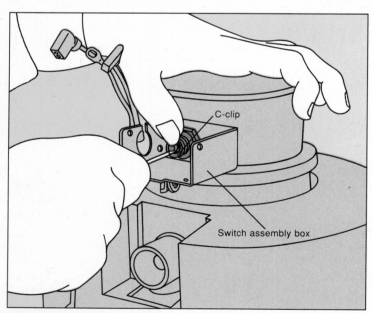

C-clip

Switch assembly box

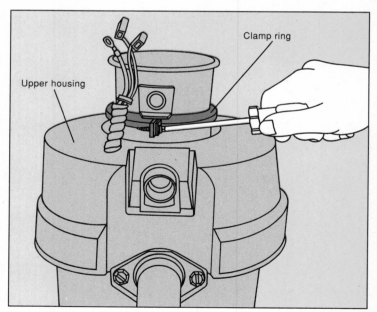

Clamp ring

Upper housing

1 **Removing the switch assembly.** Shut off power to the disposer at the main service panel; disconnect and dismount the disposer *(page 88)*. On batch-feed models, shown here, the switch assembly must first be removed to get at the upper housing. Unscrew the switch assembly cover, disconnect the terminal leads and ground wire and pull them free of the switch assembly box. Next, remove the C-clip using a small screwdriver. Hold the clip firmly; it secures a spring that, as you free it, could propel the tiny clip across the room. Remove the plunger, seal, switch box and other parts of the switch assembly, noting their positions for reassembly.

2 **Removing the upper housing.** Unscrew or unlock the clamp ring *(above)* and remove it. Remove the housing nuts or screws, then lift off the upper housing.

REPLACING THE SHREDDER RING AND FLYWHEEL (continued)

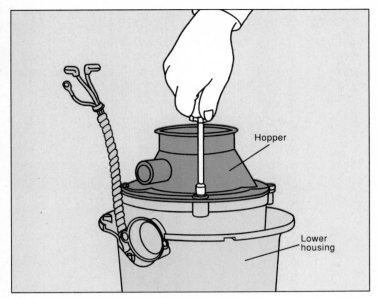

Hopper

Lower housing

3 **Removing the hopper.** Depending on your model, use a nut driver, as shown, or a screwdriver to free the hopper from the lower housing, exposing the shredder ring and flywheel.

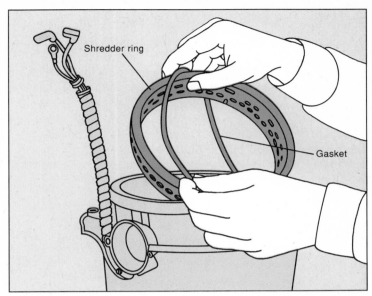

Shredder ring

Gasket

4 **Checking the shredder ring.** If the disposer grinds slowly, the shredder ring may be dull. If your model has a metal shredder pad hanging from the shredder, hold onto it as you lift out the gasket and shredder ring. Inspect the shredder ring and replace it if dull or damaged; if you can't tell whether or not the shredder ring is dull, take it to an appliance repair shop and compare its cutting edges to those of a replacement part. When you remove the shredder housing gasket, it will stretch; replace it with a new one when you reassemble the disposer.

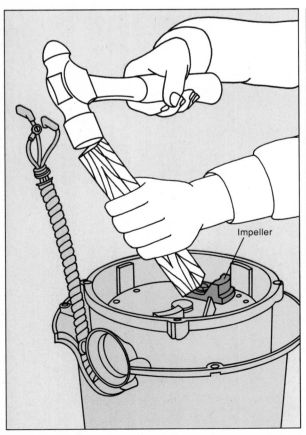

Impeller

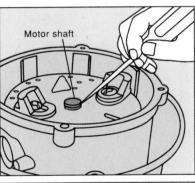

Motor shaft

5 **Replacing the flywheel.** If an impeller is broken or the flywheel is bent or damaged, replace the flywheel. On some disposers, the flywheel is threaded directly onto the motor shaft. To free it, first determine whether to drive the wheel clockwise, as shown, or counterclockwise, by tracing the shaft threads with a pointed tool *(left, top)*. Then hold a block of scrap wood against an impeller and strike the block sharply with a ball-peen hammer *(far left)*. Unwind the flywheel from the shaft.

On other disposers, the flywheel is held on the motor shaft by a nut *(left, bottom)*. Keep the wheel stationary by wedging a screwdriver against an impeller, then loosen the nut with a wrench.

Rethread the replacement flywheel onto the motor shaft or secure it to the shaft with the nut. Replace the shredder housing gasket and shredder ring, hopper, upper housing, clamp ring and switch assembly, and reinstall the disposer.

CLOTHES WASHERS

Its myriad cycles and settings make a clothes washer seem far too complex to fix. But if tackled logically and systematically, repairing a washer is well within the skills of most do-it-yourselfers.

When something does go wrong, service technicians locate and fix the trouble by dividing the washer into three systems. Then they examine one system at a time. The plumbing system brings in and circulates water through the pump and a network of hoses; the electrical system energizes the machine through the control switches, motor, timer, solenoids and their wiring; and the mechanical system—transmission, drive belt, agitator, basket and tub—powers the cleaning action.

To diagnose a problem, first determine where in the operating cycle the problem originated. Then consult the Troubleshooting Guide; its list of possible causes and repair procedures will help you pinpoint the problem and decide what to do. Do not begin a repair without referring to the pages indicated in the chart. Disassembling your washer in the hope of finding the problem is usually a waste of time.

Though all automatic washers fill, agitate, pump out and spin dry in basically the same way, there are key differences in design and special features from model to model. The two washers pictured below—referred to as Type I and Type II throughout this chapter—illustrate most of the variations found

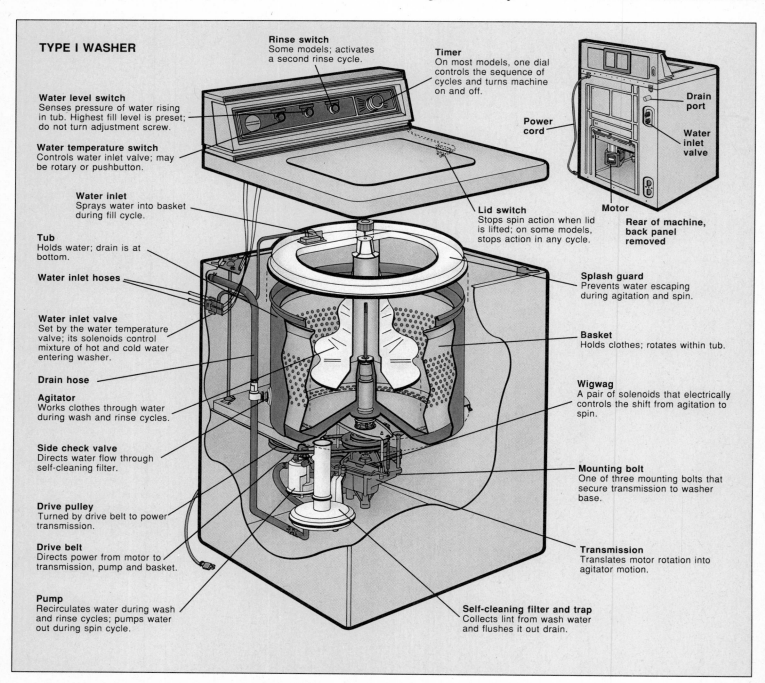

TYPE I WASHER

Rinse switch
Some models; activates a second rinse cycle.

Timer
On most models, one dial controls the sequence of cycles and turns machine on and off.

Water level switch
Senses pressure of water rising in tub. Highest fill level is preset; do not turn adjustment screw.

Water temperature switch
Controls water inlet valve; may be rotary or pushbutton.

Water inlet
Sprays water into basket during fill cycle.

Tub
Holds water; drain is at bottom.

Water inlet hoses

Water inlet valve
Set by the water temperature valve; its solenoids control mixture of hot and cold water entering washer.

Drain hose

Agitator
Works clothes through water during wash and rinse cycles.

Side check valve
Directs water flow through self-cleaning filter.

Drive pulley
Turned by drive belt to power transmission.

Drive belt
Directs power from motor to transmission, pump and basket.

Pump
Recirculates water during wash and rinse cycles; pumps water out during spin cycle.

Lid switch
Stops spin action when lid is lifted; on some models, stops action in any cycle.

Power cord

Drain port

Water inlet valve

Motor

Rear of machine, back panel removed

Splash guard
Prevents water escaping during agitation and spin.

Basket
Holds clothes; rotates within tub.

Wigwag
A pair of solenoids that electrically controls the shift from agitation to spin.

Mounting bolt
One of three mounting bolts that secure transmission to washer base.

Transmission
Translates motor rotation into agitator motion.

Self-cleaning filter and trap
Collects lint from wash water and flushes it out drain.

in modern washers. Your machine will likely resemble one of these two types. Any differences will usually be in the location of parts, rather than in their function or testing procedure. The illustrations will help you recognize a part and understand its function, even if its position is not the same in your washer.

Though similar in appearance to other top-loading washers, the direct-drive washer on page 113 has a radically different design that greatly simplifies access and repair. Front-loading tumbler washers, in spite of their lower energy costs, are not as popular as the top-loading agitator models shown here.

Most washer repairs require only wrenches, screwdrivers and pliers. To service electrical parts, you will need a continuity tester or a multitester *(page 132)*. Broken parts, including a faulty motor, are usually replaced rather than rebuilt, although repair kits are available for the pump and several other plumbing components. A faulty transmission, like its counterpart in the family car, can be professionally rebuilt.

A careful reading of your washer's Use and Care manual, available from the manufacturer, can help prevent problems caused by misuse. When repair is required, always turn off the water faucets to the machine, unplug the power cord, and have a container handy to catch water runoff. Most washers are very heavy; if you plan to lay the machine on the floor for access through the bottom, enlist a helper.

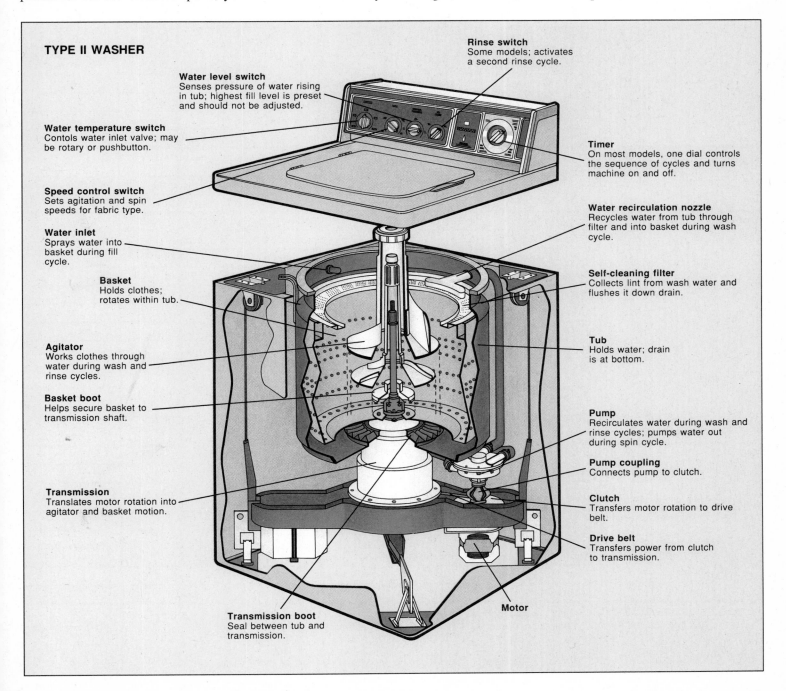

TYPE II WASHER

Rinse switch
Some models; activates a second rinse cycle.

Water level switch
Senses pressure of water rising in tub; highest fill level is preset and should not be adjusted.

Water temperature switch
Contols water inlet valve; may be rotary or pushbutton.

Timer
On most models, one dial controls the sequence of cycles and turns machine on and off.

Speed control switch
Sets agitation and spin speeds for fabric type.

Water recirculation nozzle
Recycles water from tub through filter and into basket during wash cycle.

Water inlet
Sprays water into basket during fill cycle.

Self-cleaning filter
Collects lint from wash water and flushes it down drain.

Basket
Holds clothes; rotates within tub.

Agitator
Works clothes through water during wash and rinse cycles.

Tub
Holds water; drain is at bottom.

Basket boot
Helps secure basket to transmission shaft.

Pump
Recirculates water during wash and rinse cycles; pumps water out during spin cycle.

Pump coupling
Connects pump to clutch.

Transmission
Translates motor rotation into agitator and basket motion.

Clutch
Transfers motor rotation to drive belt.

Drive belt
Transfers power from clutch to transmission.

Transmission boot
Seal between tub and transmission.

Motor

TROUBLESHOOTING GUIDE

SYMPTOM	POSSIBLE CAUSE	PROCEDURE
Washer doesn't run at all (in some cases, motor may hum)	No power to washer	Check that washer is plugged in; check for blown fuse or tripped circuit breaker *(p. 132)* □○
	Motor overheated	Turn off washer; allow motor to cool for one hour
	Lid switch faulty	Test lid switch *(p. 100)* □○
	Timer faulty	Test timer and timer motor *(p. 99)* ⬓◒▲
	Water level switch, hose or dome faulty (Type I)	Service water level switch assembly *(p. 98)* ⬓◒
	Motor start relay faulty (Type II)	Test motor start relay *(p. 112)* ⬓◒▲
	Motor faulty	Test motor *(Type I, p. 111; Type II, p. 112)* ■●▲; remove for professional service or call for service
	Pump blocked or impeller jammed	Inspect pump *(Type I, p. 105; Type II, p. 106)* ⬓●
Washer doesn't fill	Faucets turned off	Turn on faucets
	Water supply hoses kinked	Turn off faucets; disconnect and straighten hoses; replace if damaged □○
	Filter screens clogged	Clean or replace screens *(p. 103)* □○
	Water inlet valve faulty	Test water inlet valve *(p. 103)* ⬓◒▲
	Water temperature switch faulty	Test water temperature switch *(p. 100)* ⬓◒
	Water level switch, hose or dome faulty	Service water level switch assembly *(p. 98)* ⬓◒
	Timer faulty	Test timer and timer motor *(p. 99)* ⬓◒▲
Washer doesn't stop filling	Water inlet valve faulty	Unplug washer; if water continues to fill, valve is faulty; repair or replace *(p. 103)* ⬓◒▲
	Water level switch, hose or dome faulty	Service water level switch assembly *(p. 98)* ⬓◒
	Timer faulty	Unplug washer; if water stops filling, test timer *(p. 99)* ⬓◒▲
Washer doesn't agitate	Agitator worn	Inspect agitator *(Type I, p. 101; Type II, p. 102)* □○
	Drive belt faulty	Tighten or replace belt *(Type I, p. 106; Type II, p. 108)* ⬓◒
	Wigwag faulty (Type I)	Test wigwag solenoids *(p. 108)* ⬓◒▲
	Timer faulty	Test timer and timer motor *(p. 99)* ⬓◒▲
	Water level switch faulty	Test water level switch *(p. 98)* ⬓◒
	Lid switch faulty	Test lid switch *(p. 100)* □○
	Motor start switch faulty (Type I)	Test motor start switch *(p. 112)* ■●▲
	Motor start relay faulty (Type II)	Test motor start relay *(p. 112)* ⬓◒▲
	Transmission faulty	Remove transmission for professional service *(Type I, p. 109; Type II, p. 110)* ⬓◒, or call for service
	Pulley loose (Type I)	Tighten setscrew on motor pulley *(p. 111)* or transmission pulley *(p. 109)* ⬓◒
	Cam bar rivet broken (Type I)	Replace cam bar rivet with cotter pin *(p. 109)*; repair kit may be available ⬓◒
Washer doesn't drain	Drain hose too high	Check Use and Care manual; reposition drain hose
	Drain hose kinked	Straighten hose; replace if damaged
	Suds blocking drain	Turn off washer, bail out suds and hot water, add cold water to washer; use less detergent
	Tub drain blocked	Inspect tub drain *(Type I, p. 101; Type II, p. 102)* ⬓◒
	Pump blocked or impeller jammed	Inspect pump *(Type I, p. 105; Type II, p. 106)* ⬓●
	Self-cleaning filter or trap clogged (Type I)	Inspect filter and trap *(p. 104)* ⬓◒
	Timer faulty	Test timer and timer motor *(p. 99)* ⬓◒▲
	Wigwag faulty (Type I)	Test wigwag solenoids *(p. 108)* ⬓◒▲

DEGREE OF DIFFICULTY: □ **Easy** ⬓ **Moderate** ■ **Complex**
ESTIMATED TIME: ○ **Less than 1 hour** ◒ **1 to 3 hours** ● **Over 3 hours** ▲ **Multitester required**

SYMPTOM	POSSIBLE CAUSE	PROCEDURE
Washer doesn't spin	Drive belt loose or broken	Tighten or replace belt (Type I, p. 106; Type II, p. 108) ▭●
	Wigwag faulty (Type I)	Test wigwag solenoids (p. 108) ▭●▲
	Lid switch faulty	Test lid switch (p. 100) ▢○
	Timer faulty	Test timer and timer motor (p. 99) ▭●▲
	Motor start switch faulty (Type I)	Test motor start switch (p. 112) ■●▲
	Motor start relay faulty (Type II)	Test motor start relay (p. 112) ▭●▲
	Motor faulty	Test motor (Type I, p. 111; Type II, p. 112) ■●▲; remove for professional service or call for service
	Motor pulley loose (Type I)	Tighten pulley setscrew (p. 111) ▭●
	Cam bar rivet broken (Type I)	Replace cam bar rivet with cotter pin (p. 109); repair kit may be available ▭●
	Transmission faulty	Remove transmission for professional service (Type I, p. 109; Type II, p. 110) ▭●, or call for service
	Pump blocked or impeller jammed	Inspect pump (Type I, p. 105; Type II, p. 106) ▭●
	Basket drive faulty (Type I)	Call for service
	Clutch worn (Type II)	Call for service
Washer leaks	Filter clogged (models with manual filter)	Remove and clean manual filter ▢○
	Hose loose or cracked	Inspect all internal hoses; tighten hose clamps, replace damaged hoses ▢●
	Pump leaking	Inspect pump (Type I, p. 105; Type II, p. 106) ▭●
	Transmission boot faulty (Type II)	Inspect transmission boot (p. 110) ▭●
	Side check valve flapper stuck (Type I)	Call for service
	Basket drive seal faulty (Type I)	Call for service
Washer noisy or vibrates excessively	Load unbalanced	Redistribute load
	Washer not level	Adjust leveling feet
	Snubber scratching snubber plate (Type I)	Sand snubber bottom (p. 101) ▢○
	Transmission braces loose	Tighten braces (Type I, p. 106; Type II, p. 108) ▭●
	Transmission oil low (Type I)	Add oil (p. 110) ▭●
	Belt slipping	Adjust belt tension (Type I, p. 106; Type II, p. 108) ▭●
	Pump blocked or impeller jammed	Inspect pump (Type 1, p. 105; Type 11, p. 106) ▭●
	Clutch worn (Type II)	Call for service
Washer damages clothing	Cleaning agents used improperly	Check Use and Care manual
	Agitator cracked	Inspect agitator (Type I, p. 101; Type II, p. 102) ▢○
	Basket surface rough; object stuck in basket	Inspect basket (Type I, p. 101; Type II, p. 102) ▭●
Lint on clothing	Wrong mixture of fabrics in load	Check Use and Care manual
	Filter clogged (models with manual filter)	Remove and clean filter ▢○
	Tub drain blocked	Inspect tub drain (Type I, p. 101; Type II, p. 102) ▭●
	Pump blocked or impeller jammed	Inspect pump (Type I, p. 105; Type II, p. 106) ▭●
	Self-cleaning filter or trap clogged (Type I)	Inspect filter and trap (p. 104) ▭●

DEGREE OF DIFFICULTY: ▢ Easy ▭ Moderate ■ Complex
ESTIMATED TIME: ○ Less than 1 hour ● 1 to 3 hours ● Over 3 hours   ▲ Multitester required

ACCESS THROUGH THE CONTROL CONSOLE

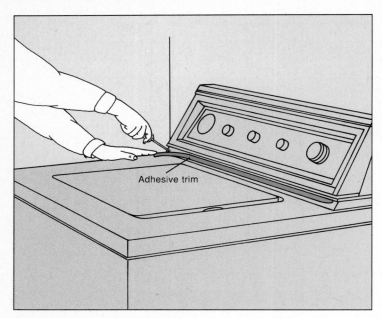

Adhesive trim

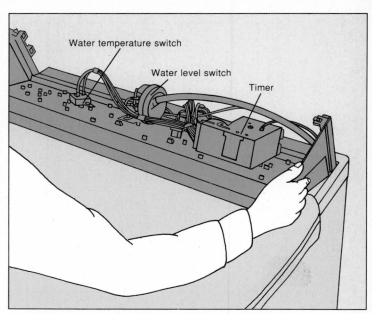

Water temperature switch

Water level switch

Timer

1 **Unscrewing the control console.** Disconnect power to the washer. To free the control console of most washers, you must remove retaining screws from the bottom front corners of the console. The screws may be hidden by a strip of adhesive trim, as shown. Other washers may have screws on the top or at the back of the console.

2 **Rolling the console panel foward.** Drape a towel over the top of the washer to protect its enamel finish. Tilt the control console forward and rest it on the washer top. If the console has a back panel, unscrew and remove it. You now have access to the timer and switches *(above)*. The wiring diagram for Type II washers is often located inside the console as well.

ACCESS THROUGH THE TOP

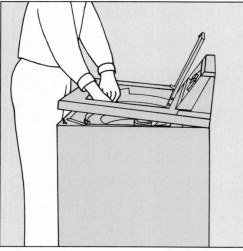

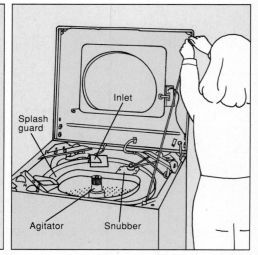

Splash guard

Inlet

Agitator

Snubber

Raising the washer top. Tape down the lid so that it will not swing open when the top is raised. On some Type II models, the recirculation nozzle under the lid must first be detached from the plastic cover shield. The washer top is held down by a spring clip near each corner. Slip a putty knife, the blade padded with masking tape, between the washer top and chassis near each corner, and push against the spring clips to release the top *(above, left)*. If the spring clips are too stiff to release easily with the putty knife, lift the lid and grasp the inner edge of the washer opening. Jerk the top forward and up to unlock the clips *(above, center)*. Lean the top against a wall, or support it with a length of heavy cord or a chain *(above, right)*. You now have access to the snubber, splash guard, inlet and agitator.

ACCESS THROUGH THE REAR PANEL

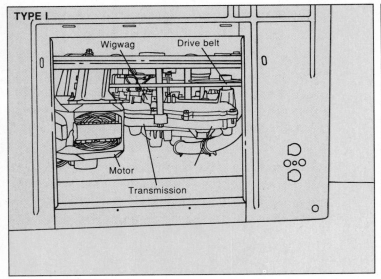

TYPE I

Wigwag   Drive belt

Motor

Transmission

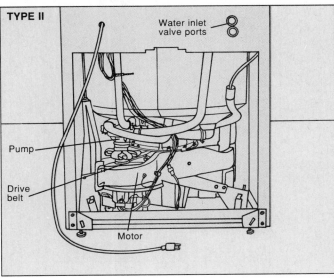

TYPE II

Water inlet
valve ports

Pump

Drive
belt

Motor

Removing the rear panel. Some washer components can be reached through an opening in the back of the chassis. Unplug the washer and pull it away from the wall. A Type II washer is very heavy; you may need help to move it. Unscrew the rear panel and set it aside. With the rear panel of a Type I washer removed *(above, left)*, you have access to the motor, drive belt, wigwag, and back filter if your model has one. You can also top up the transmission oil from this position *(page 110)*. After removing the back panel of a Type II washer *(above, right)*, you have access to the water inlet valve, pump, drive belt, motor and motor start relay.

ACCESS THROUGH THE BOTTOM

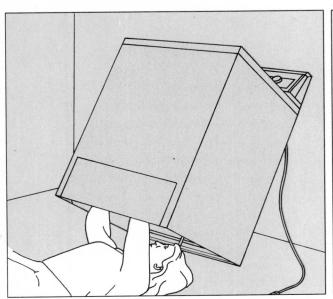

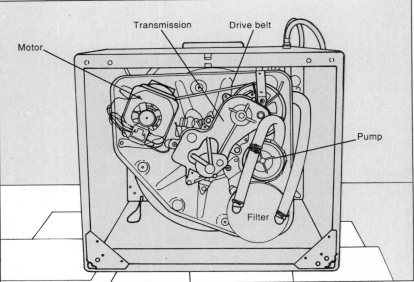

Transmission   Drive belt

Motor

Pump

Filter

Getting to parts through the bottom. Unplug the washer. Shut off the faucets and detach the water inlet hoses. Have a bucket handy to catch dripping water, and bail or siphon out any water in the tub. To perform a quick check, pull a Type I washer about 2 feet from the wall and tilt it back *(above, left)*. Be absolutely certain its position is secure, and the control console is firmly attached, before looking inside. For actual repairs, lay a Type I washer front down on the floor *(above, right)*. Protect the floor with newspaper; the transmission of a Type I washer will probably leak oil in this position. With the bottom exposed, you have access to a Type I washer's trap and filter, pump, drive belt, transmission, motor and motor start switch. On a Type II washer, removing the drive motor is the only repair performed through the washer bottom.

SERVICING THE WATER LEVEL SWITCH ASSEMBLY

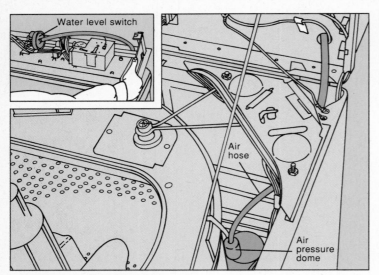

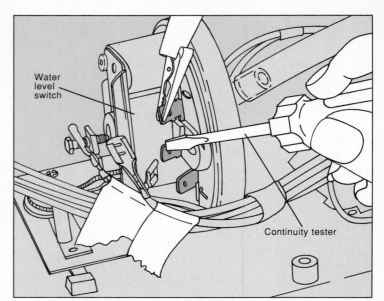

1 **Inspecting the air hose and air pressure dome.** Turn off and unplug the washer. Remove the control console *(page 96)*. Check the air hose *(inset)*; straighten any kinks and replace it if perforated. Remove the air hose from the water level switch and blow through it to clear any trapped water. Reconnect the air hose to the water level switch. Next, raise the top of the washer *(page 96)* and, on a Type I washer, follow the hose down the right side to the air pressure dome *(above)*. The connection should be completely airtight. If the dome is cracked, or if the seal between the dome and tub is broken, replace the dome. To remove it, pull off the hose, depress the dome and turn it counterclockwise one-quarter turn. Reverse the procedure to install a new dome.

2 **Testing a water level switch in the EMPTY position.** Inspect the wire connectors; if they are burned or loose, splice on new connectors *(page 136)*. Label and remove the three wires from their terminals on the water level switch. Test each terminal against the other two with a continuity tester, as shown. The tester should indicate continuity across one pair of terminals, and resistance across the other two pairs. If not, the water level switch is broken and should be replaced *(step 4)*.

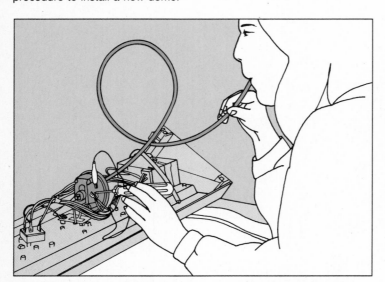

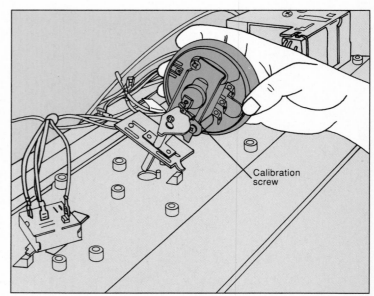

3 **Testing the water level switch in the FULL position.** After pulling the air hose off the switch's port, attach a shorter tube of the same diameter and blow very gently through it into the switch. You should hear a click when the switch trips to the FULL position. While blowing through the tube, test the terminals with a continuity tester as in step 2. This time a different pair of terminals should show continuity; the other two should show resistance. If you do not hear a click, or you have trouble blowing gently and evenly into the tube, reconnect the air hose, fill the washer with water, unplug the machine and test the switch again. If the switch fails the test, replace it *(step 4)*.

4 **Replacing the water level switch.** Pull off the water level switch control knob, if any. Remove the tube from the switch's port and unscrew the water level switch bracket from inside the control console. Do not turn the calibration screw; the slightest adjustment could result in the tub overflowing. The mounting bracket of a Type I water level switch often has a tab that fits into a slot in the control panel. Screw the bracket of the new switch inside the console, reconnect the air hose and the terminal wires, push on the control knob and replace the console.

TESTING AND REPLACING THE TIMER

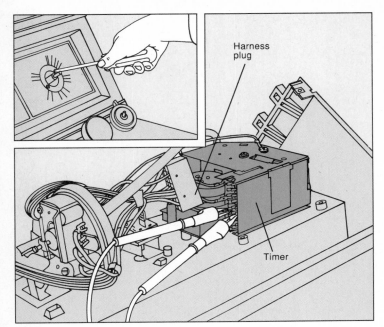

Harness plug

Timer

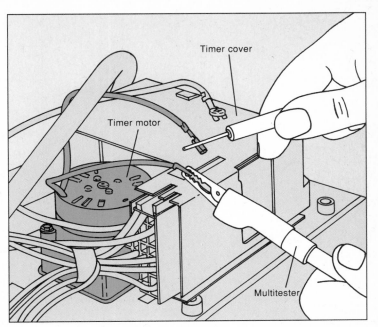

Timer cover

Timer motor

Multitester

1 **Testing and replacing the timer.** Unplug the washer and remove the control panel *(page 96)*. Check the washer's wiring diagram for the terminals that control the broken cycle, and disconnect and label those wires. (Some timers have a harness plug, as shown; just pull off the plug to expose the terminals.) Set the control knob to the affected cycle and touch the probe of a continuity tester or a multitester set at RX1 to each terminal *(above)*. The tester should show continuity; if not, replace the timer. (To test the timer motor, go to step 2.) Remove the control knob and unscrew the timer from the front *(inset)* or from the back. Screw a new timer in place, plug in the wires and replace the control knob and the console.

2 **Testing the timer motor.** Label and remove both motor wires from their terminals. Set a multitester to RX100 and touch a probe to each terminal *(above)*; the motor should produce a reading of 2,000 to 3,000 ohms. Many timer motors can be replaced separately from the timer; simply remove the two screws that hold the motor on the timer. When installing a new timer motor, make sure that its gear fits back into the hole in the timer cover.

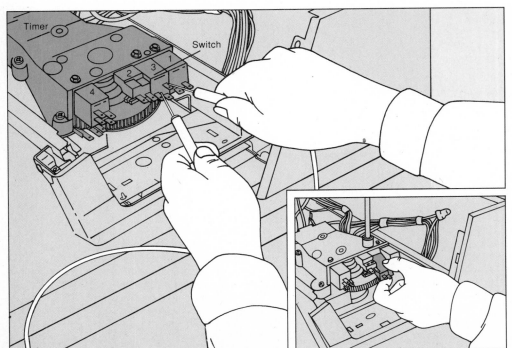

Timer

Switch

Servicing a plastic timer (Type II washers). Unplug the washer and remove the control console *(page 96)*. If your Type II washer has a plastic timer, lift off the switch cover by releasing the plastic tab. Check your wiring diagram and timer chart to determine which switch controls the broken cycle. Label and disconnect the wires from those terminals. Testing with a continuity tester or a multitester set at RX1 *(left)* should show the following results:

CONTROL KNOB POSITION	SWITCH					
	1 Motor	2 Special Function	3 Main Power	3b Bypass	4 Wash	4b Spin
OFF	open	See timer chart	open	open	open	open
WASH	open		closed	open	closed	open
SPIN	open		closed	closed	open	closed

Replace a switch that fails any test. Switches 1 and 3 are easily replaced; remove the screw that secures the switch to the plastic casing and gently pull it free from the timer *(inset)*. The other switches can only be replaced by disassembling the timer; call for service.

TESTING AND REPLACING THE WATER TEMPERATURE SWITCH

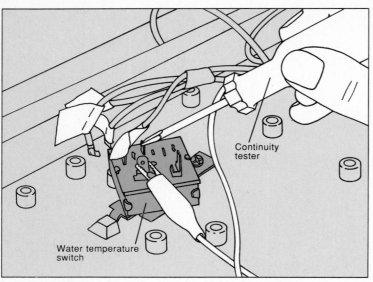

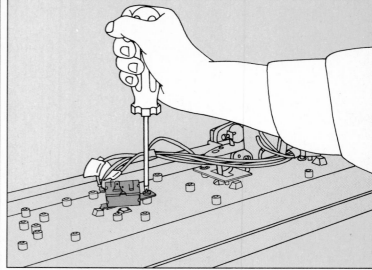

Testing the water temperature switch. Unplug the washer and remove the control console *(page 96)*. The water temperature switch may be rotary, as shown, or pushbutton; both are tested the same way. Check the wiring diagram for the markings used on the terminals that control the inoperative setting. Label and disconnect the wires from these terminals. Turn the knob to the inoperative setting or press the corresponding button. Touch one probe of a continuity tester to each terminal *(above, left)*. The tester should light; if not, replace the switch. Unscrew the old switch from the control console *(above, right)*, install the new switch and transfer the wires.

TESTING AND REPLACING THE LID SWITCH

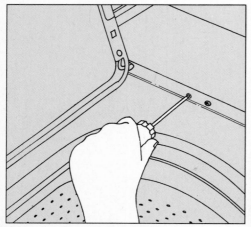

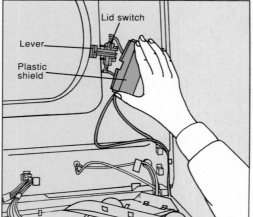

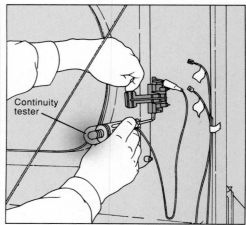

Inspecting the lid switch. Unplug the washer. To reach a Type I lid switch, loosen the two screws on the right side of the washer top *(above, left)*. Raise the top *(page 96)* and unsnap the plastic shield covering the switch *(above, center)*. Examine the switch; if both wires are secured to the terminals, the problem may be a damaged bracket or a broken switch. Lid switches vary in design. Some Type I washers have a mercury switch capsule in a bracket attached to the right lid hinge. Replace a damaged bracket by unscrewing it from the washer chassis. Type II shieldless lid switches have a metal lever; if bent, straighten it by hand. Make sure the lever covers the lid strike hole completely.

To test a switch, label and disconnect the wires. Clip a continuity tester probe to one of the terminals. Touch the second probe to the other terminal while lifting the lever with your finger *(above, right)*. The tester should light, indicating continuity. Release the lever; the continuity tester should not light. If the switch fails either test, replace it. Unscrew a Type I lid switch from inside the washer top and pull off the plastic lever. Fit the new switch in the lever and screw the switch on the washer just until the screws begin to grab. Then reconnect the wires, snap on the plastic shield, and tighten the screws completely. To replace a Type II lid switch, remove the two screws securing it to the inside edge of the washer top. Position the new switch over the lid strike hole.

SERVICING THE TUB (Type I washers)

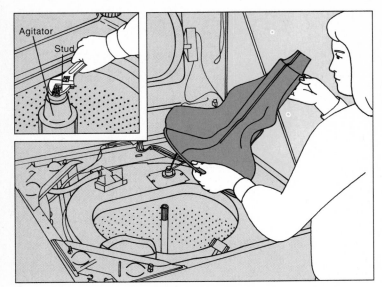

1 **Removing and inspecting the agitator.** Unplug the washer and siphon or bail out any water still in the tub. To remove the agitator from some Type I washers, you must first unscrew a plastic cap that hides a stud-and-seal assembly, then unscrew the stud with a wrench *(inset)*. On others, pry off a lid and unscrew a nut to reach the stud. Lift the agitator off the shaft *(above)*; if it sticks, tap a block of wood against it with a ball-peen hammer. Run your fingers over the agitator fins and inspect the gear teeth inside the agitator column for signs of wear. If there are cracks that might catch fabrics, or if the teeth are worn smooth, replace the agitator.

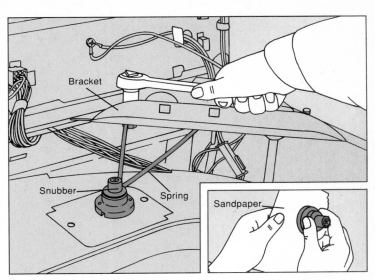

2 **Removing the snubber and spring.** To release the snubber, remove the nut and bolt securing the spring to the corner bracket *(above)*. Then pull up the spring with one hand and slip the snubber out from underneath it. If the snubber squeals, sand it *(inset)* or rub it against a rough surface such as a cement block. Free the spring arm from its hole in the bracket by twisting the spring loop forward and to the right; you may have to remove screws first. Once the arm is off the bracket, pull it through the hole.

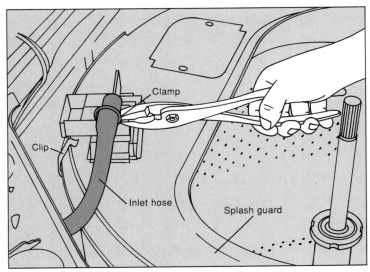

3 **Checking the water inlet hose and splash guard.** The hose and nozzle are obvious sources of leaks. Use hose-clamp pliers to unclip the spring clamp securing the inlet hose to the nozzle *(above)*. If the clamp has lost tension or is corroded, replace it with a screw-type clamp. If the nozzle is cracked, pry it off the splash guard and replace it with a new one. Replace a cracked or brittle inlet hose by pulling it from the inlet valve port behind the tub; install a new hose of exactly the same length. With the inlet hose removed, unhook the clips that secure the splash guard and lift the guard up off the tub rim. Check the gasket for wear: On some models it is glued to the under surface of the guard; on others it is fitted over the rim of the tub. Replace it if damaged.

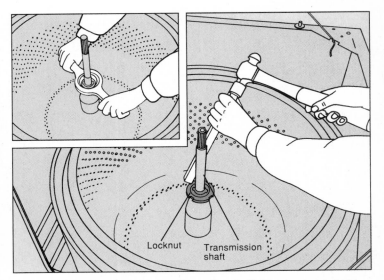

4 **Dislodging the locknut from the shaft.** The basket is held on the transmission shaft by a notched nut. To loosen it, wedge the corner of a block of wood into one of the notches, then hit the block with a ball-peen hammer to move the nut counterclockwise, as shown. The basket has a porcelain finish that can be easily chipped if you miss your mark with the hammer. If the nut is frozen to the shaft, use a spanner ring designed for your model *(inset)*, available from an appliance parts dealer.

SERVICING THE TUB (Type I washers, continued)

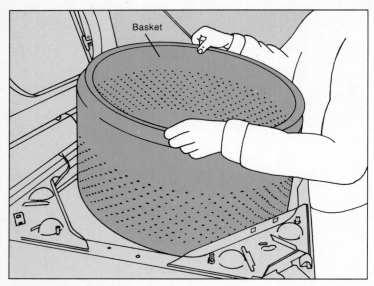

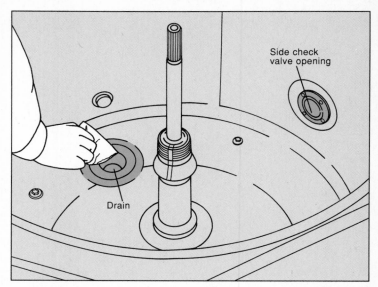

5 **Lifting out the washer basket.** Grip the basket firmly on each side, and lift it up and out of the tub in a smooth and steady motion. If the basket is rusted tight to the shaft, spray the shaft with penetrating oil and wait a few minutes, gently rock the basket back and forth, then lift. Scrub the basket inside and out with a stiff-bristled brush, then rub a folded dish towel around the inside. If you feel fabric-damaging roughness, use an emery cloth to smooth the surface. Remove any objects caught in the perforations. Replace the basket if the porcelain finish is cracked or otherwise damaged.

6 **Cleaning the tub.** Clothes (particularly baby socks or handkerchiefs) sometimes become trapped between the basket and the tub, and may clog the tub drain or the side check valve opening. After removing the basket, wipe out the bottom of the tub, as shown. Scrub the surface clean of built-up mineral or detergent deposits that might obstruct the spinning basket. Use a wire coat hanger to free hard-to-reach material blocking the outlets. If this does not clear the obstruction, disconnect the drain hose that leads from the tub to the pump and clean or replace it.

SERVICING THE TUB (Type II washers)

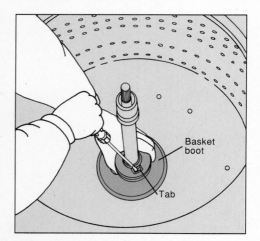

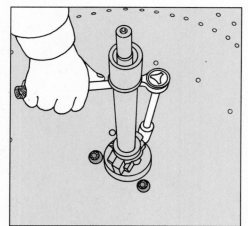

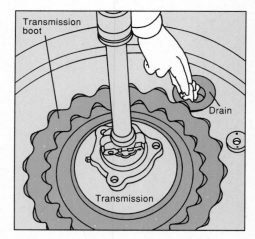

1 **Removing the agitator and basket boot.** Unplug the washer and bail out the tub. Most Type II agitators can be pulled off the shaft with a sharp tug on each side of the base. If the agitator sticks, tap it with a ball-peen hammer while pulling, or pry off the cap with a screwdriver and pour hot water or a few drops of oil on the shaft. With a screwdriver, release the spring-loaded tab on the rubber basket boot *(above)* and lift the boot up off the shaft.

2 **Unscrewing the basket.** Pull out the recirculation nozzle from the upper edge of the tub. Use a socket wrench and extension to unscrew the three bolts in the floor of the basket *(above)*. Firmly grip the lip of the basket and lift the basket straight up out of the machine. Be careful not to bang the remaining nozzles.

3 **Inspecting the drain and transmission boot.** Free the drain of obstructions *(above)* and wipe the surface of the tub clean. Make sure the transmission boot is well-seated and sealed; water leaking through the seal will damage the transmission. If the boot is cracked or brittle, replace it *(page 110)*.

SERVICING THE WATER INLET VALVE (Type I washers)

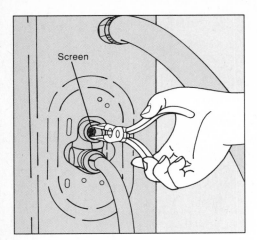

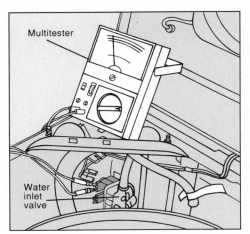

1 **Inspecting the filter screens.** Unplug the washer, turn off the faucets and disconnect the water inlet hoses at both ends. Check the screens regularly for blockage or rust. Using long-nose pliers, pry the screens out of the faucet couplings and the inlet valve ports *(above)*. Clean the screens with an old toothbrush under running water. Plastic screens can be replaced with metal ones but not vice versa.

2 **Testing the inlet valve solenoids.** Raise the top *(page 96)* and lift out the splash guard *(page 101)*. The inlet valve is in the left rear corner behind the tub. Label and remove the wires from one solenoid. Set a multitester to RX100 and clip a probe to each terminal. You should get a reading of 500-2,000 ohms. Test the second solenoid, if any, the same way. If either solenoid is faulty, replace the entire valve assembly *(step 3)*.

3 **Replacing the inlet valve.** From inside the washer, unclamp the hose from the inlet valve port, then remove the screws that secure the valve to the back of the cabinet *(above)*. Hold on to the valve with your free hand and pull the valve out through the top. Hold the new valve in position and screw it to the cabinet. Reconnect the wires and hoses.

SERVICING THE WATER INLET VALVE (Type II washers)

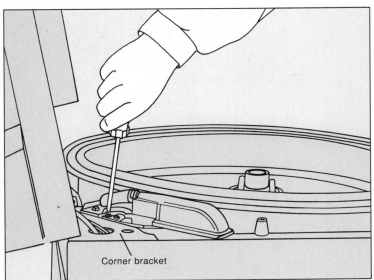

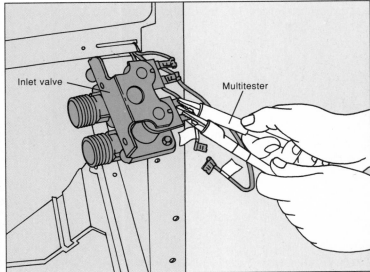

Servicing the inlet valve. Follow step 1 *(above)* to clean the filter screens. To service the valve, unplug the machine, lift the top *(page 96)* and remove the back panel *(page 97)*. Remove the screws holding the inlet valve to the corner bracket *(above, left)* and the washer cabinet. Then push the inlet valve down so that you can test it through the back of the machine *(above, right)*. Label and disconnect the wires. Set the multitester at RX100 and touch a probe to each terminal of one solenoid; the tester needle should sweep partially upscale, indicating resistance. Test the second solenoid the same way. If the inlet valve is faulty, remove it by unscrewing the clamp securing the hose to the inlet port. Install a new valve by reversing this procedure.

SERVICING THE SELF-CLEANING FILTER AND TRAP (Type I washers)

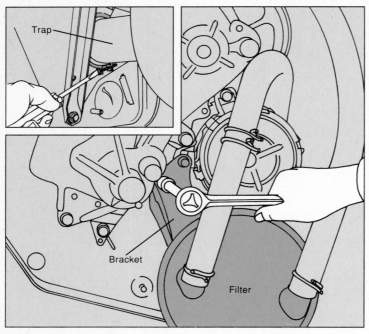

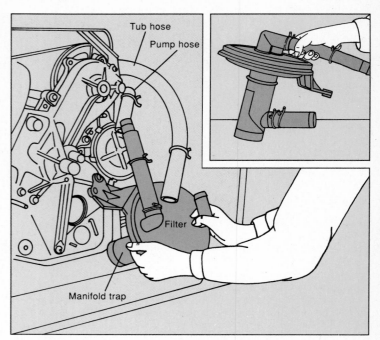

1 **Releasing the filter and trap.** Some Type I washers have self-cleaning filters connected to the tub outlet through a manifold trap. To remove a clogged filter and trap assembly, unplug the washer, turn off the water faucets and detach the hoses. Have a bucket handy to catch dripping water. Bail or siphon out any water remaining in the tub. Lay the washer flat on some newspapers *(page 97)*. Unscrew the clamp securing the pipe-like manifold trap to the tub outlet *(inset)*. Use a socket wrench to remove the two bolts that hold the plastic filter bracket to the transmission, as shown.

2 **Removing and cleaning the trap.** The filter ports are clamped to a long hose leading to the tub, and a short hose leading to the pump. Disconnect the long hose from its port on the filter, and disconnect the short hose from its port on the pump. As you pull the filter and trap assembly free *(above)*, notice its proper position for reinstallation. Check the filter and manifold trap for cracks or for debris that may block the flow of water. Tap the end of the trap on the floor to knock out any objects caught inside *(inset)*; pins and buttons are common culprits. If the filter is clogged, replace it. To replace the filter and trap assembly, reverse this procedure.

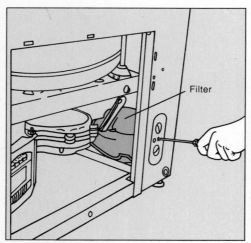

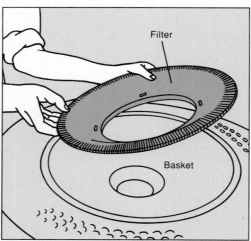

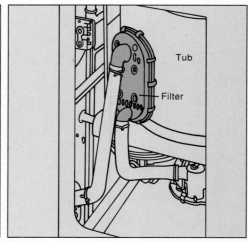

Alternate filters. Some older Type I washers have a self-cleaning filter in the lower right rear corner of the cabinet *(above, left)*. To replace it, remove the single screw that secures it to the cabinet and use pliers to unclamp the hoses. Other Type I models use a ring filter *(above, center)* that snaps on the bottom of the basket. To remove it, take out the basket *(page 102)*, turn it over and use a screwdriver to pry off the four clips securing the ring filter to the base of the basket. More recent models have a tub-mounted filter attached with a special locknut that also holds the side check valve in place *(above, right)*. The drain hose exits through the lower left rear corner of the cabinet. Replacement of this filter is difficult and rarely required. If it must be removed, call for service .

REPLACING THE PUMP (Type I washers)

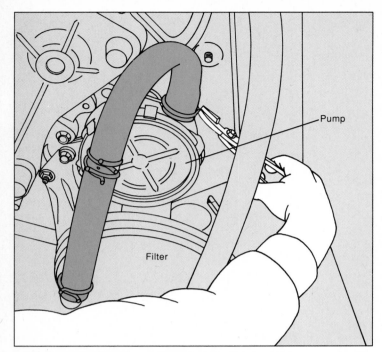

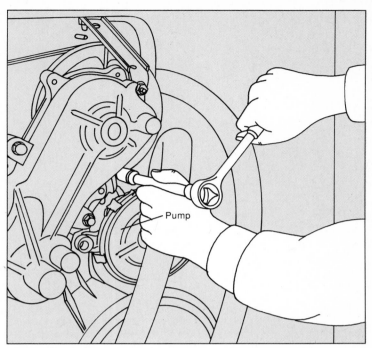

1 **Unclamping the pump hoses.** A pump may have two, three or four ports, and may be made of metal or plastic. Most pumps are designed to be thrown out when broken. To remove a pump, you must disconnect the hoses. Unplug the washer, turn off the faucets and detach the hoses. Have a bucket handy to catch dripping water. Bail or siphon out any water still in the washer basket. Lay the washer flat on newspaper *(page 97)*. Washers with self-cleaning filters usually have two-port pumps; one hose connects to the filter, the other to the trap. Use hose-clamp pliers to loosen the clamps on the two hoses *(above)* and pull the hoses off the ports.

2 **Unscrewing the bolts.** Use a socket wrench with an extension to reach the two pump mounting bolts *(above)*. Sometimes one of the bolts is shared with the filter. After releasing the pump, replace this bolt. As you pull the pump free, notice for replacement purposes how the control lever on top of the pump engages in the slot in the agitator cam bar on the transmission. Inspect the pump for damage. Look into the ports for any visible obstruction. When you move the lever, the flapper valve inside should open and close the ports tightly. Spin the pump pulley. A wobbly pulley, broken flapper valve or cracked plastic shell are all indications that you need a new pump.

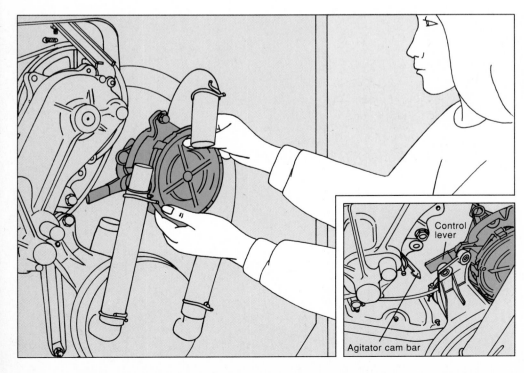

3 **Installing a new pump.** Position the pump so that the control lever slips into the slot of the agitator cam bar on the transmission *(inset)*. Bolt the pump to the transmission. Work the hoses back onto the ports and reposition the clamps.

REPLACING THE PUMP (Type II washers)

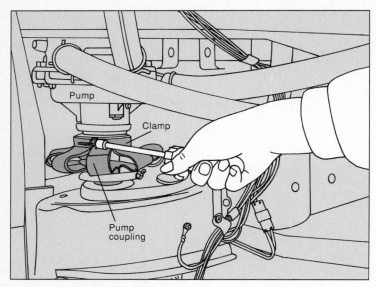

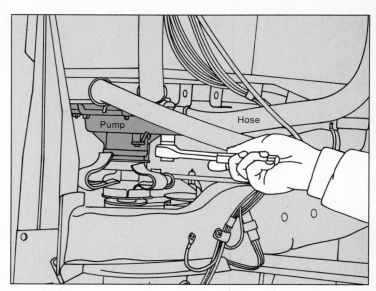

1 **Removing the screw clamp from the pump coupling.**
Unplug the washer. Turn off the faucets and detach the hoses. Ready a bucket to catch water runoff, and bail or siphon any water in the tub. Pull the washer out from the wall and remove the rear access panel *(page 97)*. Unscrew the clamp securing the pump coupling to the pump, as shown.

2 **Dismounting the pump.** Use a socket wrench to unscrew the three screws that hold the pump to the tub *(above)* and lift out the pump. Slide the hose clamps off the pump ports and pull the hoses free. Inspect the pump. Impellers jammed by foreign objects can sometimes be released by pulling the object out through a port. Damaged impellers, cracks or leaking seals mean the pump must be discarded. Install an exact replacement, reattach the hoses, tighten the hose clamps, and secure the pump to the coupling.

ADJUSTING AND REPLACING THE DRIVE BELT (Type I washers)

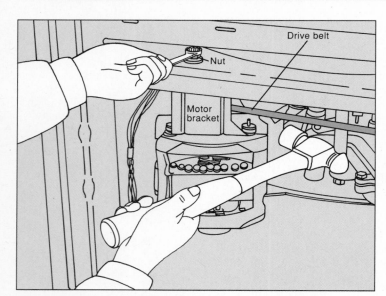

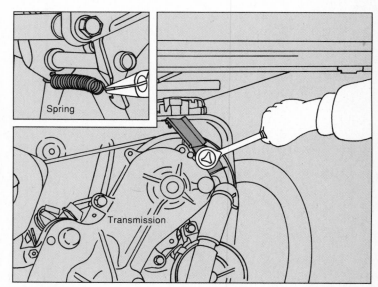

1 **Adjusting the drive belt.** Unplug the machine and turn off the faucets. Pull the washer away from the wall and remove the back panel *(page 97)*. Press the drive belt with your thumb; it should not deflect more than 1/2 inch. To tighten the belt, loosen the motor bracket nut just enough so that the motor can be shifted slightly. Tap the right edge of the drive motor bracket with a ball-peen hammer, as shown, to tighten the belt. When the belt tension feels correct, retighten the nut. If the belt is worn, proceed to step 2.

2 **Removing the transmission support braces.** To take out a worn belt and install a new one, the transmission must be pulled away slightly from the washer base. Unplug the washer and lay it on the floor *(page 97)*. Use a socket wrench to unscrew the five nuts and one bolt that hold the braces to the transmission *(above)*. Unbolt and remove the pump *(page 105)*. Use long-nose pliers to unclip the lower end of the spring stretching between the clutch plate and the transmission *(inset)*.

ADJUSTING AND REPLACING THE DRIVE BELT (Type I washers, continued)

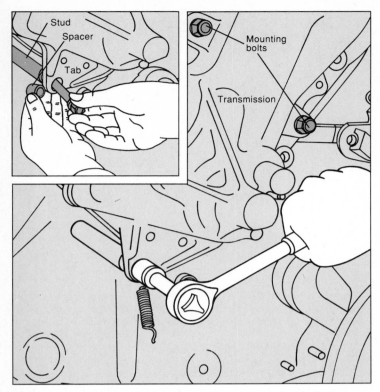

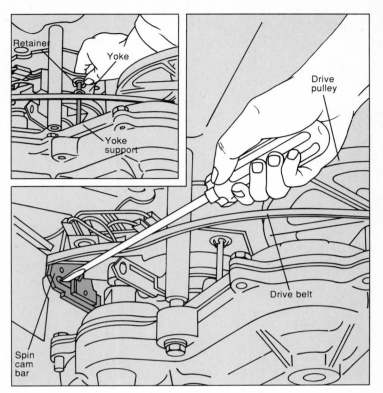

3 **Unscrewing the three transmission mounting bolts.** Loosen the two mounting bolts on each side of the transmission about 1/2 inch (7 to 10 turns). Then completely remove the lower bolt, as shown. As you pull out the bolt, catch the spacer that will fall from between the chassis stud and the transmission tab *(inset)*. Slip the old drive belt through the space.

4 **Shifting the clutch shaft.** Working through the opening in the back of the washer, use a screwdriver to pull up on the spin cam bar *(above)* while turning the main drive pulley with your hand. The movement of the spin cam bar will allow the clutch shaft to drop a bit, making space for the drive belt to pass between the shaft and the yoke *(step 5)*. Do not lose the clutch shaft washers. Snap the rod-like yoke support out of the plastic retainer on the yoke *(inset)*.

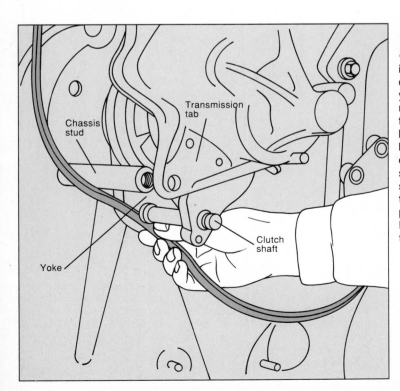

5 **Removing the drive belt.** Reaching through the bottom of the washer, pull the transmission straight toward you until it stops against the mounting bolts. Pass the old drive belt between the clutch shaft and the yoke, as shown, and remove it from the machine. To install a new belt, pass it between the chassis stud and transmission tab, then between the clutch shaft and yoke. Reassemble the washer, carefully reversing the procedures in steps 2 through 5. Reposition the yoke support and clutch shaft, replace the spacer, screw in the three mounting bolts, and replace the transmission support braces. Reinstall the pump. Loop the drive belt around the four pulleys, ending with the drive pulley. Finally, check the belt tension *(step 1)*.

ADJUSTING AND REPLACING THE DRIVE BELT (Type II washers)

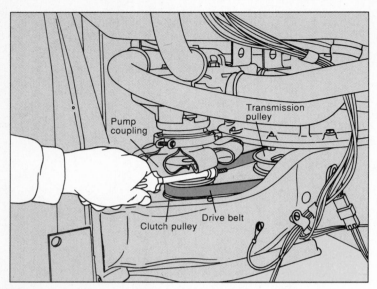

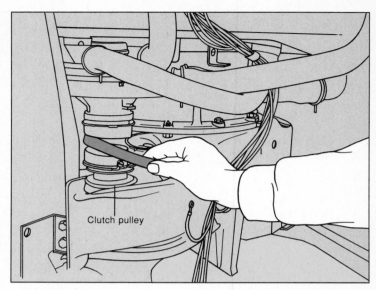

1 **Removing the drive belt.** Unplug the washer and take off the back panel *(page 97)*. To adjust the belt, see step 2. To remove it, unscrew the clamps above and below the pump coupling, as shown, and remove the coupling. Using a socket wrench with an extension, loosen the three motor mounting nuts on the mounting plate to release tension on the belt. Reach under the tub and turn the transmission pulley as you pry off the belt, then remove the belt from the clutch pulley. Pull the belt out of the machine.

2 **Installing and adjusting the new belt.** Fit the belt around the transmission pulley and hold it there while you loop it around the clutch pulley *(above)*. If the belt is tight, rotate the transmission pulley and shift the motor slightly to the right. Reinstall the pump coupling and tighten the clamps. Push the belt with your thumb; if you can deflect it more than 1/2 inch, pull the motor toward you to tighten it, then tighten the motor mounting nuts.

TESTING AND REPLACING THE WIGWAG (Type I washers)

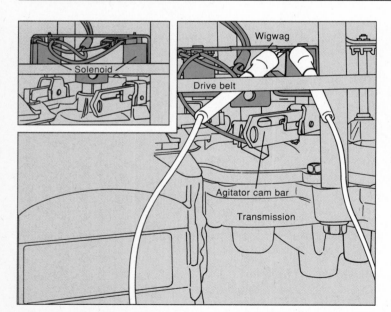

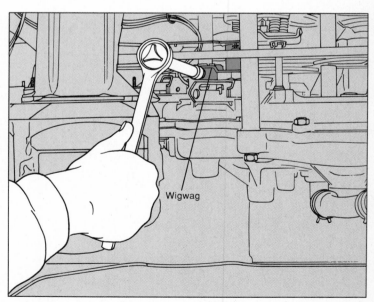

1 **Testing the wigwag.** Unplug the washer and remove the back panel *(page 97)*. The wigwag—a pair of solenoids—sits on top of the transmission. The left solenoid controls the spin cam bar and the right one controls the agitator cam bar *(inset)*. To test either solenoid, label and remove its two wires. With a multitester set at RX10, touch a probe to each solenoid terminal, as shown. The multitester should show about 200 to 700 ohms. No continuity means the solenoid is faulty; replace the wigwag.

2 **Replacing the wigwag.** Label and disconnect the wires. Using a socket wrench with an extension, remove the tapered screw from the front of the wigwag, as shown, and lift the wigwag out of the washer. Install the new wigwag in the same position. When reconnecting the terminal wires, make sure they pass through the top bracket of the wigwag.

SERVICING THE TRANSMISSION ASSEMBLY (Type I washers)

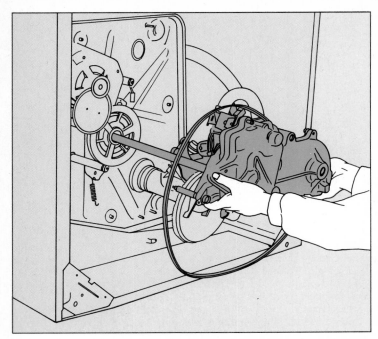

1 **Removing the transmission assembly.** Unplug the washer and remove the agitator *(page 101)*. Follow the instructions for replacing the drive belt *(pages 106 and 107)*, but remove all three transmission mounting bolts *(page 107, step 3)*. Label and disconnect the wigwag terminal wires *(page 108)*. Now pull the transmission assembly straight out toward you, as shown. If you intend to replace the transmission or have it serviced professionally, next remove the external components.

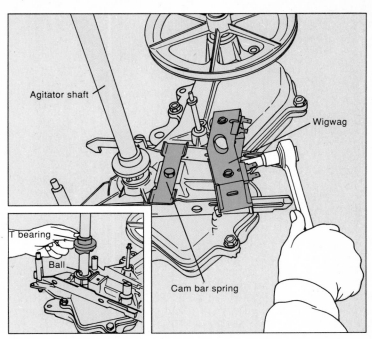

2 **Removing the external components.** Using a socket wrench with an extension, first unscrew the wigwag, as shown, then the cam bar spring. Slide the T bearing up and off the agitator shaft *(inset)*; slip the ball out of its hole at the base of the shaft.

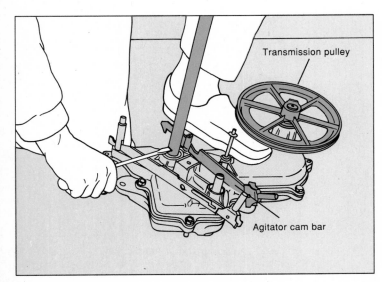

3 **Freeing the agitator cam bar and the transmission pulley.** To release the agitator cam bar, insert the end of a screwdriver into the hole in the base of the agitator from which you removed the ball in step 2. Using the screwdriver as a lever, lift the agitator shaft, as shown. With the shaft raised, pull the agitator cam bar free of its retaining fork. To remove the transmission pulley, use a hex wrench to loosen the setscrew on its hub, and lift off the pulley.

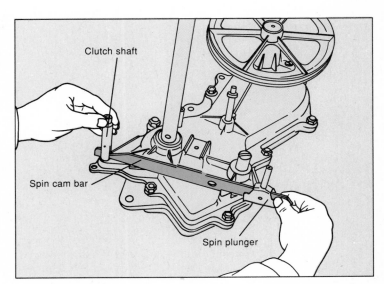

4 **Removing the spin cam bar.** Lift the clutch shaft and slide out the spin cam bar, as shown. Inspect the plunger on each cam bar. If the rivet that holds it in place is worn or broken, replace it with a cotter pin. To reinstall the transmission assembly, carefully follow in reverse the procedures described in steps 1 through 4.

SERVICING THE TRANSMISSION ASSEMBLY (Type I washers, continued)

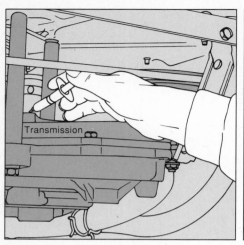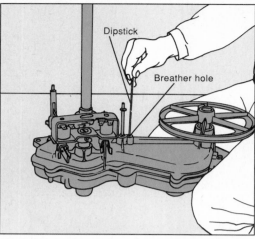

Refilling the transmission with oil. When the washer lies flat on the floor for a period of time, a small amount of oil will leak out of the transmission. After setting the washer upright, use a syringe to inject non-detergent, SAE-90 oil into the breather hole near the top center of the transmission *(above, left)*. If the washer makes a heavy clunking sound when operating, the transmission may be low on oil. To check

the oil level, remove the transmission from the washer *(page 109)*. Insert a straightened wire hanger into the breather hole *(above, center)*. Make sure the dipstick hits the bottom of the transmission and not the connecting rod between the gears. The oil should reach 3/4 inch up the dipstick. If not, use a plastic squeeze bottle to add more oil *(above, right)* until the proper level is reached.

REPLACING THE TRANSMISSION (Type II washers)

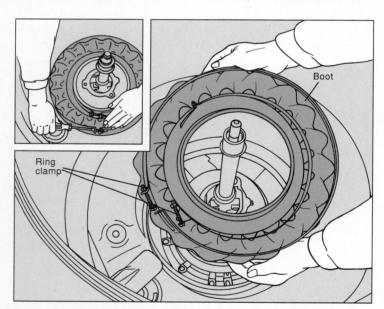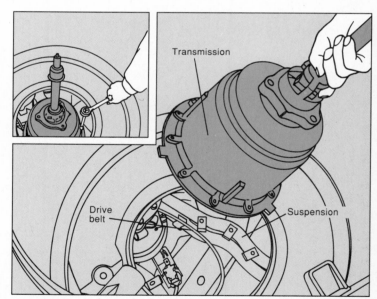

1 **Removing the transmission boot.** Unplug the washer and take out the basket *(page 102)*. Use a nut driver to loosen the two metal ring clamps that hold the boot in position *(inset)*. Then grasp the edges of the boot and pull up sharply to break the seal. Lift the boot and rings out of the tub, as shown. Wipe the clamp rings clean. Examine the boot for any damage that may allow water to leak through to the transmission and floor. Replace it if necessary.

2 **Removing and replacing the transmission.** Use a socket wrench with an extension to unscrew the six bolts anchoring the transmission to the washer's suspension *(inset)*. Noting its position for reinstallation, lift the transmission up and out of the washer *(above)*. A faulty transmission must be serviced professionally. To replace the transmission, drop one bolt into its right rear hole, then lower the transmission carefully onto the suspension. Screw in the bolt loosely and, using it as a pivot, swing the transmission clockwise, leaving a space. Slip your hand under the transmission and fit the drive belt on the transmission pulley. Pivot the transmission back into position and screw in the five remaining bolts securely. Reinstall the transmission boot and basket.

SERVICING THE MOTOR (Type I washers)

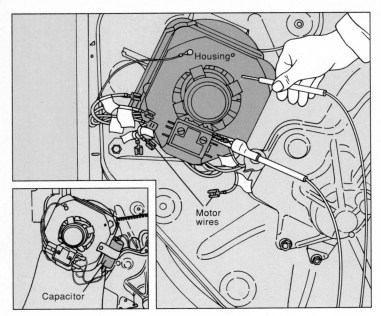

Motor wires

Housing

Capacitor

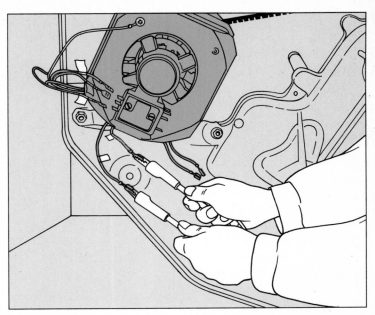

1 **Testing for a ground.** Unplug the washer and lay it on its front *(page 97)*. If there is an external capacitor on the motor housing *(inset)* discharge it first by placing one lead of a 20-ohm, wire-wound resistor on each capacitor terminal. To test the motor for grounding, set a multitester at RX1000. Touch one probe to the bare metal of the motor housing, and touch the other to each wire connector in turn. The tester needle should not move. If the motor fails this test, replace it *(steps 3 and 4)*.

2 **Testing the motor.** Most Type I washers, have a two-speed motor. One-speed and three-speed motors have a different number of wires, but all motors are tested the same way. Label and disconnect all the motor wires. Set a multitester at RX1 and clip one probe to the white wire. Touch the other probe to each colored wire in succession. In each case the multitester should show a low resistance, between 1 and 20 ohms. If you do not get a correct reading, replace the motor *(step 3)*. If the motor tests OK, next test the motor start switch *(step 5)*.

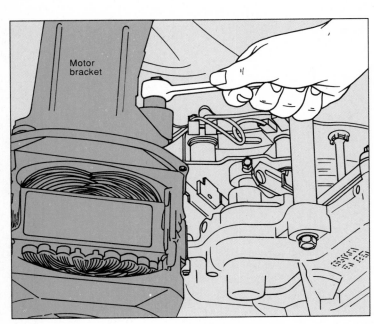

Motor bracket

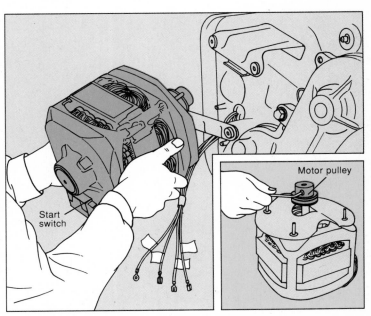

Start switch

Motor pulley

3 **Freeing the motor from its support bracket.** If the motor has an external capacitor, make sure it is discharged *(step 1)*. Label and disconnect all wires to the motor and motor start switch. Working through the back of the washer, use a socket wrench to remove the four nuts that hold the motor to the motor brackets *(above)*. Disengage the drive belt from the motor pulley; if the belt is too tight, loosen the two motor-bracket nuts on the washer base and shift the motor.

4 **Disconnecting the motor components.** Grasp the motor firmly in both hands and pull it out carefully *(above)*. Label and disconnect any wires to the capacitor and the start switch and unscrew them. Unscrew the motor pulley setscrew and remove the pulley *(inset)*. If you intend to replace the motor, save all parts for reinstallation on the new motor. Install a new motor by reversing the procedures in steps 3 and 4. Be sure to reattach the green ground wire to the motor housing.

SERVICING THE MOTOR (Type I washers, continued)

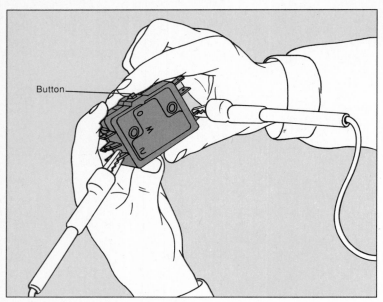

5 **Testing the motor start switch.** The motor start switch terminals are coded with colors, numbers or letters that correspond to the motor-wire colors. A typical two-speed motor uses the colors described below; consult the washer's wiring diagram if your model has different designations.

Label and remove the wires from the start switch and unscrew the switch from the motor. To test the switch, place the probes of a multitester set at RX1 (or a continuity tester) on the pairs of terminals that correspond to the wire colors listed in the chart below, or indicated in your wiring diagram. Test each pair of terminals with the switch button in, then out. The tester should show resistance or continuity according to the chart; if not, replace the start switch. Screw the new switch to the motor and reconnect the wires.

	BLUE/BLACK	ORANGE/BLACK	ORANGE/VIOLET
OUT	Resistance	Resistance	Continuity
IN	Continuity	Continuity	Resistance

SERVICING THE MOTOR (Type II washers)

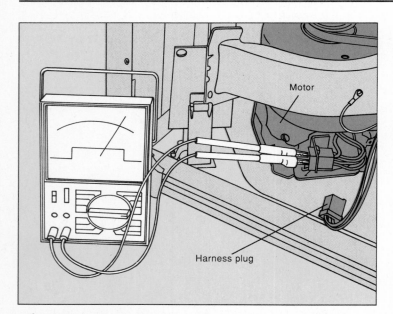

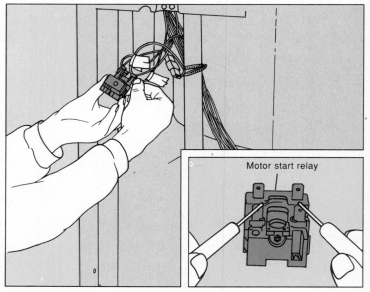

1 **Testing the motor.** Unplug the washer and remove the back panel *(page 97)*. Pull apart the plastic harness plug that houses the motor wire connections. Set a multitester to RX1 and touch the probes first to the terminals of the black and orange wires, and then to the terminals of the blue and yellow wires. The tester should indicate continuity. Set the multitester at RX1000 and test the motor for grounding by touching one probe to the bare metal of the motor housing and the other probe to each terminal in turn; there should be no continuity. To remove the motor, unclamp the lower end of the pump coupling *(page 108)*. Use a socket wrench to unscrew the three nuts securing the motor to the mounting plate. Slip the drive belt from the clutch pulley *(page 108)* and lower the motor and clutch to the floor. With a helper, tilt the washer forward and slide the motor out. If the clutch is faulty, call for service. If the motor is faulty, replace it and the start relay *(step 2)* by reversing the steps here and on page 108.

2 **Testing and replacing the motor start relay.** The start relay is mounted on the upper left corner of the washer chassis, behind the back panel. Unscrew the relay and label and disconnect its wires *(above)*. Position the relay with the arrow pointing up. Set a multitester at RX1 and touch the probes to the terminals marked M and LS; the tester should indicate continuity. Then touch the probes to terminals L and S; the tester should show resistance. Without removing the probes, turn the relay upside down. You should hear the contacts click and the tester needle should sweep upscale, indicating continuity. Replace the motor start relay if it or the motor is faulty; screw the new relay to the chassis and reconnect the wires.

DIRECT-DRIVE CLOTHES WASHER

A unique design makes the recently introduced direct-drive washer very easy to service. Diagnosis and repair of most problems involving switches, valves, agitator, basket and tub are the same as for a Type I washer. However, difficult repairs for hard-to-reach components—motor, pump, transmission and drive belt—are greatly simplified or eliminated.

In a departure from most washers, the direct-drive model has no drive belt between the motor, transmission and pump. This eliminates replacing a worn or broken belt, one of the most time-consuming repair chores, as well as reducing the number of parts that can break down.

The direct-drive washer never needs to be pulled out from the wall or lowered onto the floor for service. Access to the washer's internal parts is provided by a cabinet that pulls off and lifts away, and a control console that flips back to reveal the control switches, timer and capacitor. For convenient inspection and testing, the pump and motor are mounted under the tub in the front of the machine, or at one side. For replacement of either part, the retaining clips securing the pump and motor can be snapped free, the faulty part lifted out and a new part snapped in place.

Most direct-drive washers are equipped with a self-cleaning filter; others have a manual filter mounted on the agitator. To clean a manual filter, pull up the top of the agitator and clean the filter screen by hand.

This type of washer can also save water and detergent: Instead of automatically pumping dirty wash water down the drain, it can be modified to pump the water into a storage tub. There the suspended dirt particles will settle to the bottom; the water is then recycled in the next wash.

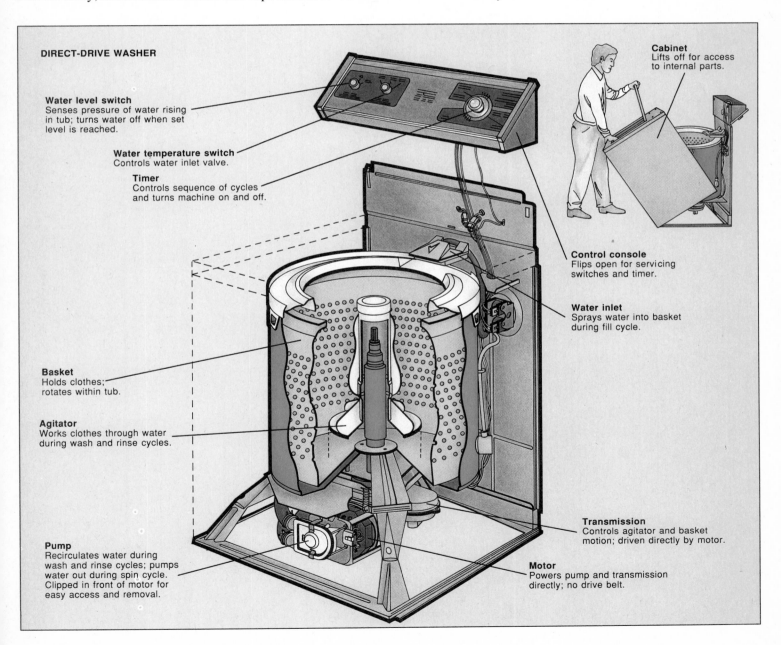

DIRECT-DRIVE WASHER

Water level switch
Senses pressure of water rising in tub; turns water off when set level is reached.

Water temperature switch
Controls water inlet valve.

Timer
Controls sequence of cycles and turns machine on and off.

Cabinet
Lifts off for access to internal parts.

Control console
Flips open for servicing switches and timer.

Water inlet
Sprays water into basket during fill cycle.

Basket
Holds clothes; rotates within tub.

Agitator
Works clothes through water during wash and rinse cycles.

Pump
Recirculates water during wash and rinse cycles; pumps water out during spin cycle. Clipped in front of motor for easy access and removal.

Transmission
Controls agitator and basket motion; driven directly by motor.

Motor
Powers pump and transmission directly; no drive belt.

CLOTHES DRYERS

A dryer combines air, heat and motion to dry everything from soggy socks to damp dungarees. The motor turns a drive belt that revolves the drum. At the same time the blower, also powered by the motor, forces air past electric heater coils (or the flame from a gas burner) and into the drum. The air draws lint and moisture from the clothes through the lint screen and out the exhaust duct. Electric switches and the timer regulate the drying time and cycles, and thermostats control the temperature. Dryers have a long life expectancy—15 years or more—and fortunately, when things do go wrong, they are as simple as they are sturdy, making most repairs worth tackling yourself.

Most electric dryers in the U.S. and Canada resemble Type I *(below)*; another popular model is Type II *(page 115)*. A gas dryer substitutes a gas burner for the heater element or coils in either machine. Many common repairs, such as replacing the timer, are similar for all dryers. When repairs are significantly different—for example, replacing the heater element, coils or gas burner—they are shown separately. Follow the instructions for the type of dryer that most resembles yours.

When a problem occurs, first check your home's main service panel for blown fuses or tripped circuit breakers *(page 132)*. Remember that an electric dryer runs on 240-volt

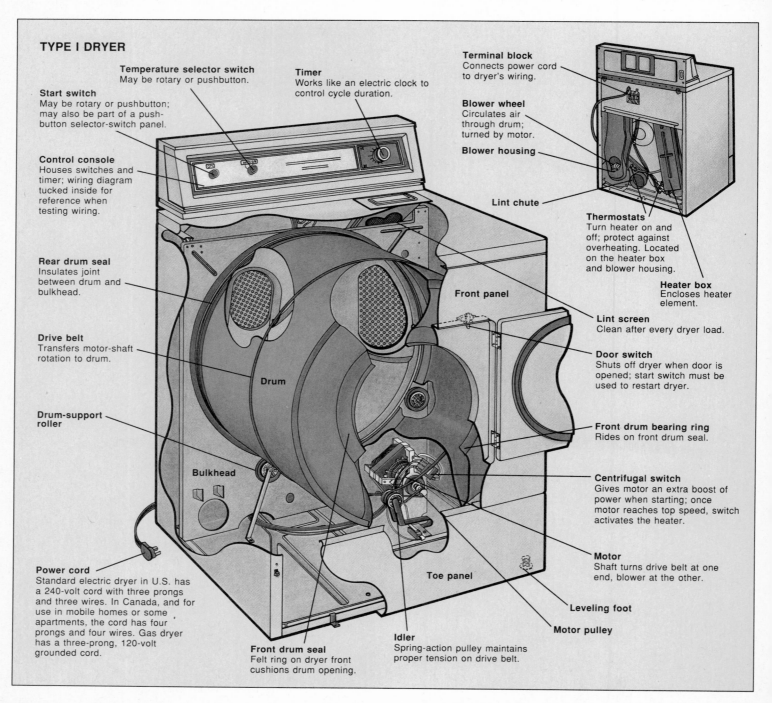

TYPE I DRYER

Start switch
May be rotary or pushbutton; may also be part of a push-button selector-switch panel.

Temperature selector switch
May be rotary or pushbutton.

Timer
Works like an electric clock to control cycle duration.

Terminal block
Connects power cord to dryer's wiring.

Blower wheel
Circulates air through drum; turned by motor.

Blower housing

Control console
Houses switches and timer; wiring diagram tucked inside for reference when testing wiring.

Lint chute

Thermostats
Turn heater on and off; protect against overheating. Located on the heater box and blower housing.

Rear drum seal
Insulates joint between drum and bulkhead.

Front panel

Heater box
Encloses heater element.

Drive belt
Transfers motor-shaft rotation to drum.

Lint screen
Clean after every dryer load.

Door switch
Shuts off dryer when door is opened; start switch must be used to restart dryer.

Drum

Drum-support roller

Front drum bearing ring
Rides on front drum seal.

Bulkhead

Centrifugal switch
Gives motor an extra boost of power when starting; once motor reaches top speed, switch activates the heater.

Motor
Shaft turns drive belt at one end, blower at the other.

Power cord
Standard electric dryer in U.S. has a 240-volt cord with three prongs and three wires. In Canada, and for use in mobile homes or some apartments, the cord has four prongs and four wires. Gas dryer has a three-prong, 120-volt grounded cord.

Toe panel

Leveling foot

Motor pulley

Front drum seal
Felt ring on dryer front cushions drum opening.

Idler
Spring-action pulley maintains proper tension on drive belt.

current and draws power through two separate fuses or breakers. As a result, if only one fuse has blown or breaker has tripped, the dryer motor may still run, although the dryer won't heat.

A dryer's greatest enemy is lint. Even if the lint filter is cleaned after every load, lint will still accumulate around the moving parts of a dryer as well as in the exhaust duct and vent, forcing the machine to work harder. At least once a year, turn off the power, remove the front and rear panels *(page 118)*, and vacuum or brush out lint from around the motor, idler and gas burner, if any. Disconnect the exhaust duct and remove lint from the internal exhaust pipe, the duct and the vent. Make sure

the duct has no kinks where lint and moisture can accumulate. The duct should be made of aluminum; plastic ducts soften and sag with heat. It must have no more than two bends of 90 degrees, spaced at least 4 feet apart.

Before working on the dryer, always unplug the power cord from the wall outlet. If the cord is wired directly into your home's electrical system, cut power to the machine at the main service panel. If you have a gas dryer, don't risk rupturing the gas line by moving the dryer or disconnecting the gas line yourself. Call the gas company or a service technician to disconnect and move it for you.

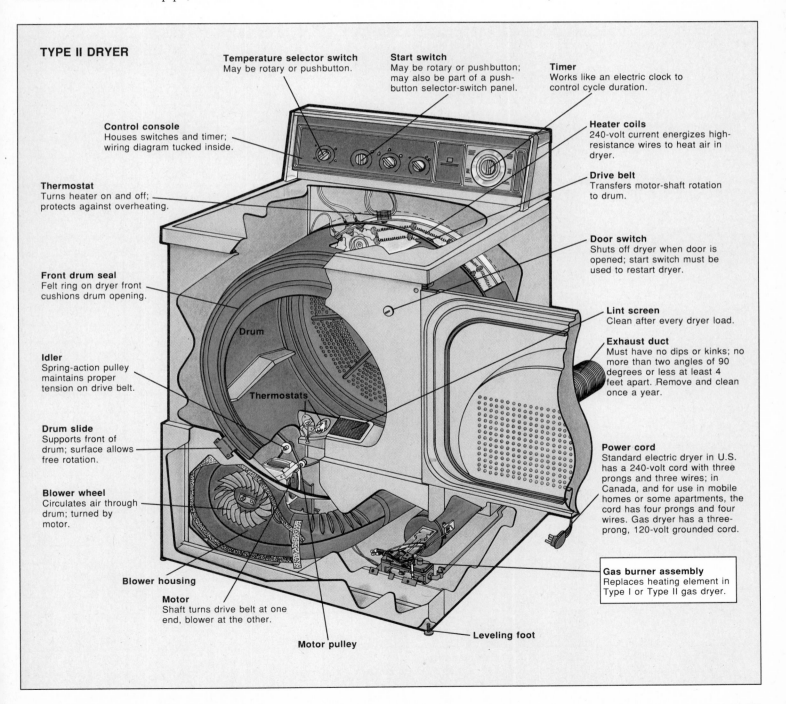

TYPE II DRYER

Temperature selector switch
May be rotary or pushbutton.

Start switch
May be rotary or pushbutton; may also be part of a push-button selector-switch panel.

Timer
Works like an electric clock to control cycle duration.

Control console
Houses switches and timer; wiring diagram tucked inside.

Heater coils
240-volt current energizes high-resistance wires to heat air in dryer.

Thermostat
Turns heater on and off; protects against overheating.

Drive belt
Transfers motor-shaft rotation to drum.

Door switch
Shuts off dryer when door is opened; start switch must be used to restart dryer.

Front drum seal
Felt ring on dryer front cushions drum opening.

Lint screen
Clean after every dryer load.

Exhaust duct
Must have no dips or kinks; no more than two angles of 90 degrees or less at least 4 feet apart. Remove and clean once a year.

Idler
Spring-action pulley maintains proper tension on drive belt.

Drum

Thermostats

Drum slide
Supports front of drum; surface allows free rotation.

Power cord
Standard electric dryer in U.S. has a 240-volt cord with three prongs and three wires; in Canada, and for use in mobile homes or some apartments, the cord has four prongs and four wires. Gas dryer has a three-prong, 120-volt grounded cord.

Blower wheel
Circulates air through drum; turned by motor.

Blower housing

Motor
Shaft turns drive belt at one end, blower at the other.

Gas burner assembly
Replaces heating element in Type I or Type II gas dryer.

Motor pulley

Leveling foot

TROUBLESHOOTING GUIDE

SYMPTOM	POSSIBLE CAUSE	PROCEDURE
Dryer doesn't run at all	No power to dryer	Check that dryer is plugged in; check for blown fuse or tripped circuit breaker *(p. 132)* □○
	Door switch faulty	Test door switch *(p. 120)* ▣○
	Start switch faulty	Test start switch *(p. 119)* ▣○
	Timer faulty	Test timer and timer motor *(p. 119)* ▣○▲
	Centrifugal switch faulty	Test centrifugal switch *(p. 119)* ▣○
	Thermostat faulty or thermal fuse blown (Type I)	Test thermostats and thermal fuse *(p. 120)* ▣○
	Power cord loose or faulty	Test power cord *(p. 133)* ▣○
	Terminal block burned	Inspect power cord terminal block *(p. 135)* ▣○
	Motor faulty	Test motor *(p. 127)* ■●▲; remove for professional service, or call for service
Motor runs, but dryer doesn't heat	Fuse blown or circuit breaker tripped	Check for blown fuse or tripped circuit breaker *(p. 132)* □○
	Temperature selector switch faulty	Test temperature selector switch *(p. 119)* ▣○
	Timer faulty	Test timer and timer motor *(p. 119)* ▣○▲
	Thermostats faulty	Test thermostats *(p. 120)* ▣○
	Centrifugal switch faulty	Test centrifugal switch *(p. 119)* ▣○
	Heater element faulty (Type I)	Test heater element *(p. 124)* ▣○▲
	Heater coils faulty (Type II)	Test heater coils *(p. 125)* ▣●▲
	Gas burner faulty (gas dryers)	Test ignitor and flame detector *(p. 126)* ▣○▲; if OK, take gas burner assembly for professional service, or call for service
Motor runs, but drum doesn't turn	Drive belt worn or broken	Check drive belt *(p. 121)* ▣○
	Idler faulty	Check idler *(p. 121)* ▣○
	Drum is binding	Service drum *(p. 122)* ▣●
Dryer runs with door open	Door switch faulty	Test door switch *(p. 120)* ▣○
Dryer doesn't turn off	Room too cool	Room must be at least 50°F for dryer to work properly
	Timer faulty	Test timer and timer motor *(p. 119)* ▣○▲
	Thermostats faulty	Test thermostats *(p. 120)* ▣○
	Heater element faulty (Type I)	Test heater element *(p. 124)* ▣○
	Heater coils faulty (Type II)	Test heater coils *(p. 125)* ▣●▲
Drying time too long	Lint screen full or exhaust duct blocked	Clean lint screen, exhaust duct and vent
	Thermostats faulty	Test thermostats *(p. 120)* ▣○
	Heater element faulty (Type I)	Test heater element *(p. 124)* ▣○▲
	Heater coils faulty (Type II)	Test heater coils *(p. 125)* ▣●▲
	Gas burner faulty (gas dryers)	Test ignitor and flame detector *(p. 126)* ▣○▲; if OK, take gas burner assembly for professional service, or call for service
Drying temperature too hot; clothes overheat	Exhaust duct or vent blocked	Clean or unkink exhaust duct; clean exhaust vent
	Thermostats faulty	Test thermostats *(p. 120)* ▣○
	Heater element grounded (Type I)	Test heater element *(p. 124)* ▣○▲
	Heater coils grounded (Type II)	Test heater coils *(p. 125)* ▣●▲
Dryer noisy	Dryer not level	Adjust leveling feet
	Loose part, panel or trim	Tighten screws on loose part, panel or trim
	Drive belt worn	Check drive belt *(p. 121)* ▣○
	Idler worn or broken	Check idler *(p. 121)* ▣○
	Object in drum seal	Check drum seals *(Type I and II, p. 122; Type I, p. 123)* ▣○
	Support rollers worn (Type I)	Check drum support rollers *(p. 122)* ▣○
	Drum shaft bearing worn (Type II)	Check drum bearing *(p. 122)* ▣○
	Blower loose or obstructed	Service blower *(Type I, p. 127; Type II, p. 128)* ▣○

DEGREE OF DIFFICULTY: □ **Easy** ▣ **Moderate** ■ **Complex**
ESTIMATED TIME: ○ **Less than 1 hour** ○ **1 to 3 hours** ● **Over 3 hours**

▲ **Multitester required**

ACCESS THROUGH THE CONTROL CONSOLE

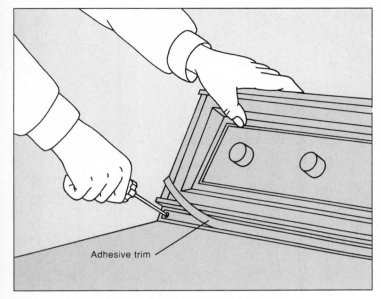

Adhesive trim

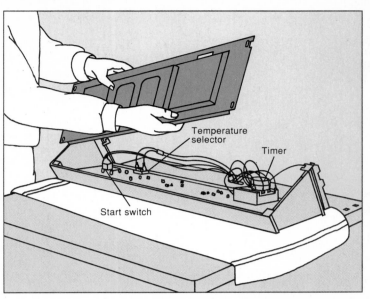

Temperature selector

Timer

Start switch

1 **Freeing the control console.** After unplugging the dryer, unscrew the control console at each end. On many machines, the screws are located at the bottom front of the console, and may be covered by adhesive trim, as shown. On other machines, the screws are located at the top or sides of the console.

2 **Removing the console back panel.** Spread a towel on top of the dryer to protect its finish. Roll the console facedown onto the towel; on some dryers, you must first slide the console forward to disengage tabs on the end panels from slots in the dryer top. If the console has a rear panel, unscrew it to expose the start switch, temperature selector, circuit diagram and timer.

ACCESS THROUGH THE TOP

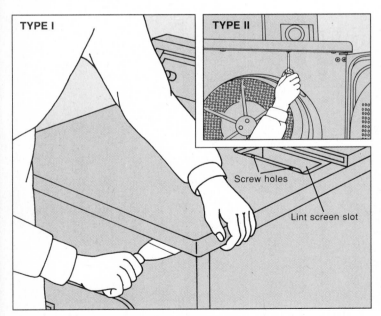

TYPE I

TYPE II

Screw holes

Lint screen slot

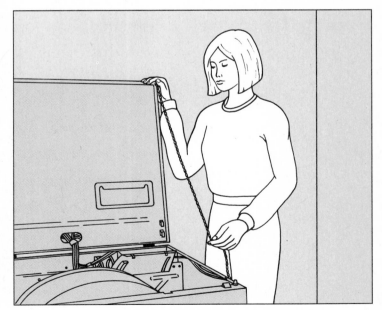

1 **Raising the dryer top.** For Type I dryers, which have a top-mounted lint screen, unplug the dryer, pull out the screen, and remove the two screws at the front edge of the screen slot. Then insert a putty knife wrapped in masking tape under the top, as shown, about two inches from each corner, and push in to disengage the hidden clips securing the top. For Type II dryers, unplug the dryer, open the door and remove the row of screws beneath the front edge of the dryer top *(inset)*.

2 **Securing the top.** The top of both models is hinged at the back; raise it and lean it against the wall behind the dryer. If the dryer is pulled out from the wall, attach a chain or cord to the top and cabinet to keep it from falling backward and damaging wiring.

ACCESS THROUGH THE REAR PANEL AND TOE PANEL

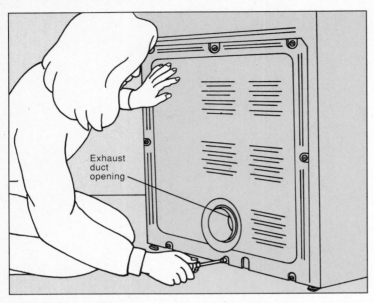

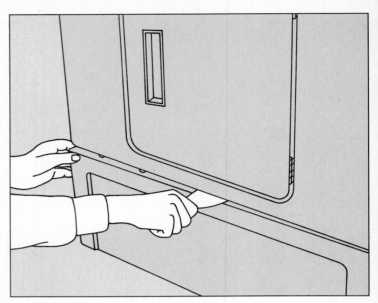

Removing the rear panel. Unless you have a gas dryer, unplug the machine, disconnect the exhaust duct and move the dryer out from the wall. (**Caution:** If you have a gas dryer, *do not* move the dryer yourself—call the gas company or a service technician to move it for you.) Most dryers have one large rear panel; remove the screws around its edges and set it aside. Some models have two or three small panels; remove each as needed for a particular repair.

Removing the toe panel. Unplug the dryer, remove any retaining screws, and insert a putty knife near the center top of the toe panel. Push down and in against the hidden clip while pulling the panel at one corner. Lift the panel off the two bottom brackets.

ACCESS THROUGH THE FRONT PANEL

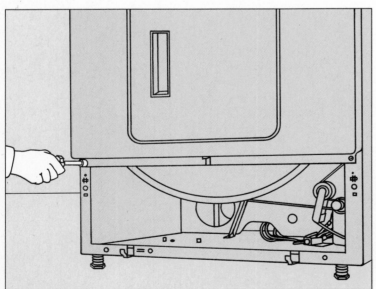

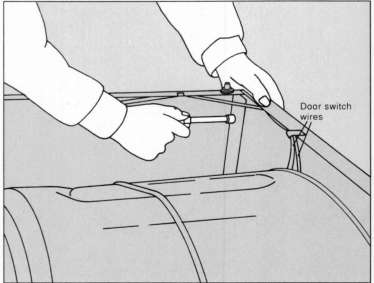

Removing the front panel. Unplug the dryer, raise the dryer top *(page 117)* and remove the toe panel, if any. (If when you remove the toe panel, you see door-hinge springs, tape the top of the door shut with masking tape to keep it from falling open, and unhook the springs from the brackets at the bottom front of the dryer before removing the front panel.) For a Type I dryer, slip a length of scrap wood under the drum to keep it from falling when you remove the front panel.

Then, for both models, loosen but do not remove the screws, if any, at the bottom corners of the front panel *(above, left)*. Dryers without front panel screws have hidden brackets inside the machine. Move to the inside of the dryer and, taking care to label their positions, disconnect the wires leading to the door switch. Supporting the front panel with one hand, remove the screws at each inside corner *(above, right)*. Lift the panel off the lower screws or brackets.

TESTING AND REPLACING SWITCHES

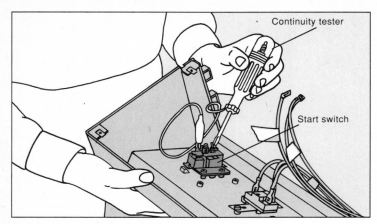

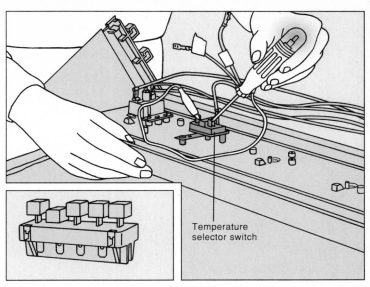

Testing the start switch. After unplugging the dryer, free the control console and tilt it forward *(page 117)*. Disconnect the wires from the start switch terminals and label their positions with masking tape. To test a two-terminal switch, place one probe of a continuity tester on terminal CO (or R2), and the other on NO (or R1), as shown. The tester should not light. Press the start button; the tester should now light, showing continuity. To test a three-terminal switch, place one probe on terminal NC (or CT1), and the other on CO (or R1); the tester should light. Press the start button; the tester should not light. If the switch fails this test, replace it. Pull off the control knob and unscrew the switch from the console. Remove and reuse the mounting bracket if the switch has one. Screw the new switch in place and reconnect the wires. If your dryer does not use these terminal configurations, consult the wiring diagram *(page 138)*.

Testing and replacing the temperature selector switch. The selector switch may be rotary or pushbutton *(inset)*; both are tested the same way. Unplug the dryer and free the control console *(page 117)*. Check the dryer's wiring diagram *(page 138)* for the markings used on the terminals regulating the inoperative cycle. Disconnect the wires from these terminals and label them. Turn the knob to the inoperative cycle or press the corresponding button. Touch one probe of the continuity tester to each terminal, as shown. The tester should light; if not, replace the switch. Unscrew the old switch from the control console, install the new switch and transfer the wires.

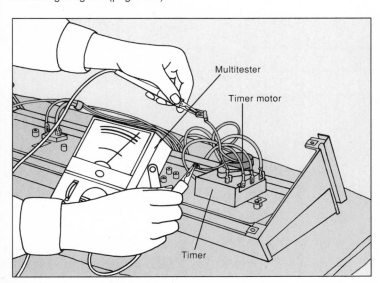

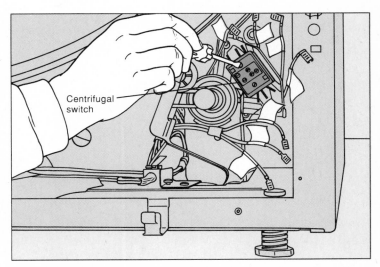

Testing the timer. Unplug the dryer and free the control console *(page 117)*. To test the timer motor, disconnect its two black wires. Set a multitester to the RX1000 scale and connect a probe to each motor terminal, as shown. The meter should show 2,000 to 3,000 ohms. If not, replace the motor by removing the two screws holding it to the timer and screwing a new motor in place. Reconnect the wires.

To test the timer itself, check the dryer's wiring diagram *(page 138)* for the configuration of the affected cycle and disconnect the wires. Set the timer knob to the cycle. Touch one probe to each terminal; the needle should swing, indicating continuity. If the timer fails this test, replace it. Pull the control knob off the front (if the timer has a mounting bracket, reuse it). Install a new timer and transfer the wires from the old timer to the new one.

Testing the centrifugal switch. Mounted on the motor, the centrifugal switch is reached in most Type I dryers either by removing the toe panel *(page 118)*; or by raising the top *(page 117)* and removing the front panel *(page 118)*. In Type II models, the centrifugal switch is reached through the rear panel *(page 118)*. Disconnect and label the wires, then unscrew the switch from the motor *(above)*. Place the probes of a continuity tester on terminals 1 and 2, then 5 and 6, then 5 and BK (or 3). Test with the switch button in, then out. Replace the switch if it does not show the following results:

	1 — 2	5 — 6	5 — BK(3)
OUT	Continuity	Continuity	Resistance
IN	Resistance	Resistance	Continuity

TESTING AND REPLACING SWITCHES (continued)

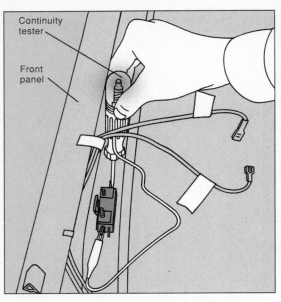

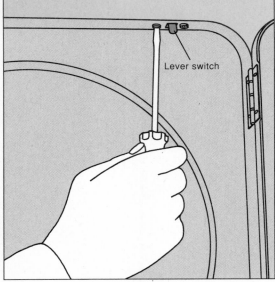

Testing and replacing the door switch. Raise the top *(page 117)* to reach the switch, which is mounted near one of the upper corners of the dryer front. Disconnect the wires from the terminals. If the switch has two wires, touch a continuity tester probe to each terminal that was connected to a wire *(above, left)*; ignore any extra terminal. With the door closed, the tester should light, showing continuity; with the door open, it should not light.

In a dryer with a drum light, the door switch will have three wires and three terminals. Clip one probe to the common terminal (at one end of the switch or, on a cylindrical switch, the largest of the three) and touch the other probe to each of the other terminals in turn. With the door closed, the tester should light with one terminal and not light with the other; with the door open, the situation should reverse.

To replace a faulty lever-style switch *(above, center)*, remove the screws on either side of the lever. Lift out the switch from inside the dryer. To remove a cylindrical switch *(right, top)*, reach down inside the dryer, squeeze the retainer clips on the back of the switch and pull it out through the front. To remove a hinge-mounted switch *(right, bottom)*, take off the dryer front panel *(page 118)* and unscrew the switch from the door hinge.

TESTING AND REPLACING THERMOSTATS

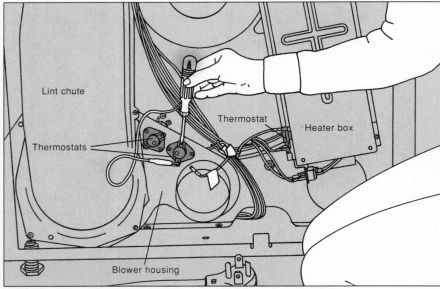

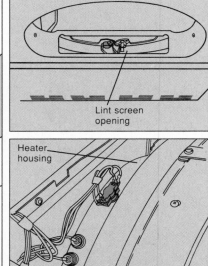

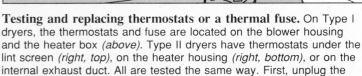

Testing and replacing thermostats or a thermal fuse. On Type I dryers, the thermostats and fuse are located on the blower housing and the heater box *(above)*. Type II dryers have thermostats under the lint screen *(right, top)*, on the heater housing *(right, bottom)*, or on the internal exhaust duct. All are tested the same way. First, unplug the dryer. Disconnect the wires from the thermostat or fuse terminals and label their positions. Touch a continuity tester probe to each terminal, as shown; the tester should light. If either of the thermostats or the fuse fails this test, replace the faulty part. Reconnect the wires and replace the dryer panels.

SERVICING THE DRIVE BELT AND IDLER

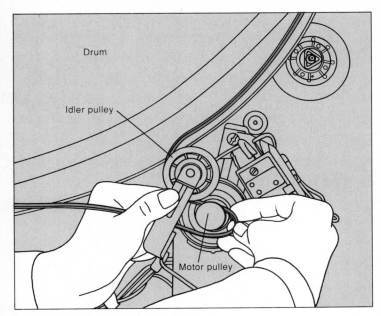

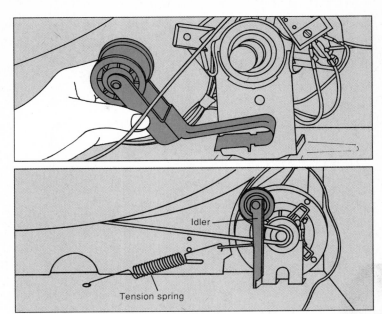

1 **Disengaging the drive belt.** To reach the idler on a Type I dryer, remove the toe panel *(page 118)* or raise the top *(page 117)* and remove the front panel *(page 118)*. Prop the dryer drum on a piece of scrap wood. Push the idler pulley toward the motor pulley, releasing tension on the drive belt, and slip the belt off the motor pulley *(above)*. To access the drive belt in a Type II dryer, remove the rear panel *(page 118)*. Pull the idler pulley away from the motor pulley to release tension on the belt and slip it off the motor pulley.

2 **Removing the idler.** With the belt disengaged, inspect the idler bracket, pulley and spring. Idlers vary in style; many are one piece and are held in place in the dryer floor by belt tension *(above, top)*. Lift the idler out and check the pulley for uneven wear or wobbling. If it is damaged, install a new idler. Another type of idler has a tension spring *(above, bottom)*. Unhook the spring and replace it if worn or broken. This type of idler may also have a replaceable pulley; go to step 3 to inspect and replace it.

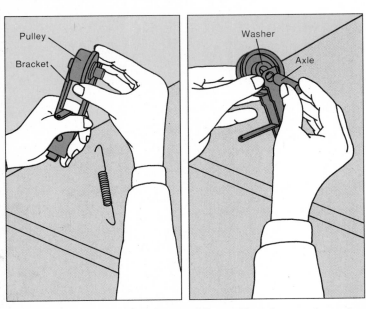

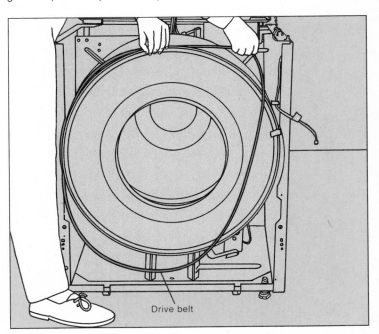

3 **Checking and replacing the idler pulley.** Inspect the surface of the pulley for uneven wear, and move it back and forth to check for wobbling *(above, left)*. To replace the pulley, use a nut driver to remove the screw at one end of the axle and slide the axle out of the pulley. Place a new pulley and washers in the bracket *(above, right)*, insert the axle and replace the screw. Some idler pulleys have a retainer ring instead of a screw, and the axle is permanently connected to the bracket. Pry off the ring with long-nose pliers to remove and replace the pulley; snap the retainer ring back on the axle. When replacing the idler, engage the spring on the idler bracket before threading the drive belt around it.

4 **Removing and replacing the drive belt.** Raise the dryer top *(page 117)* and remove the front panel *(page 118)*. Lifting the drum slightly, slide the loose belt free. Align a new belt in the same position as the old one, its grooved side against the drum. To rethread the belt in a Type I dryer, push a loop of the belt under the idler pulley and catch it on the motor pulley. Check that the rear drum seal rides properly on the bulkhead *(page 123)*. To rethread a Type II belt, loop it over the idler pulley and under the motor pulley. Turn the drum by hand to make sure the belt is properly positioned.

SERVICING THE DRUM

TYPE I

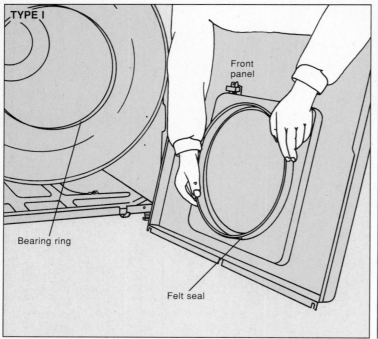

Front panel

Bearing ring

Felt seal

TYPE II

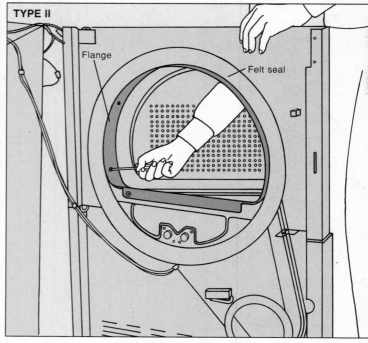

Flange

Felt seal

1 **Checking and replacing the front drum seal.** Unplug the dryer, lift the dryer top *(page 117)*, and remove the toe panel, if any, and the front panel *(page 118)*. Inspect the felt seal surrounding the door opening behind the front panel, and look for objects embedded in the felt. To replace the seal in a Type I dryer *(above, left)*, first peel off or unclip the old seal. Place a new seal with its folded

edge toward you and fit the holes in the seal over the clips on the rim of the door opening. Also check the plastic bearing ring within the drum opening; if it is rough or worn, snap it out and replace it. To replace a Type II seal *(above, right)*, unscrew the metal flange on which the felt is mounted, and screw on a new seal-and-flange assembly.

TYPE I

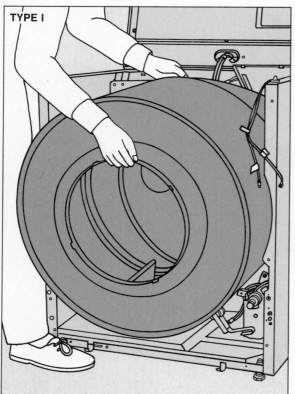

TYPE II

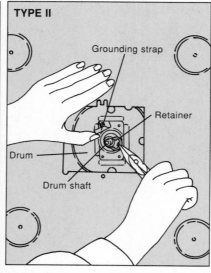

Grounding strap

Retainer

Drum

Drum shaft

2 **Removing the drum.** Disengage and remove the drive belt *(page 121)*. For a Type I dryer, lift the drum slightly and carefully slide it out through the front of the cabinet *(far left)*. For a Type II dryer, first unscrew the rear access panel *(page 118)* to expose the drum shaft. Loosen the grounding strap using a nut driver, then pry the retainer off the shaft with long-nose pliers *(near left)*. Carefully lift the drum out through the front of the cabinet.

SERVICING THE DRUM (continued)

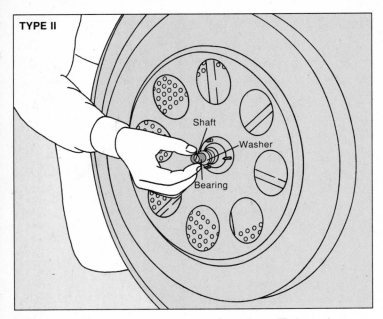

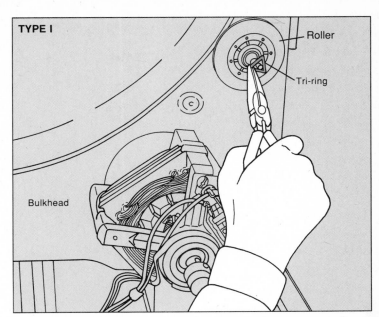

3 **Replacing the drum shaft bearing (Type II dryers).** A flexible, sleeve-like bearing cushions the drum shaft of a Type II dryer. Pull off the bearing, as shown, and inspect it for wear. If it needs to be replaced, slide a new bearing onto the shaft, making sure the washers are in their proper positions. When reinstalling the drum *(step 6)*, be sure the bearing and washers do not slip off the drum shaft.

4 **Replacing drum support rollers (Type I dryers).** A pair of rubber rollers mounted on the bulkhead supports the drum in Type I dryers. With the drum removed, check each roller for wear and replace if damaged. To remove a roller, use long-nose pliers to pry off the tri-ring that secures it to the shaft, and slide the roller off. (If the left roller has a support bracket, first unscrew it from the shaft.) Lightly lubricate the shaft with machine oil, slide on a new roller, and pop the tri-ring back onto the shaft. Screw the bracket back in place.

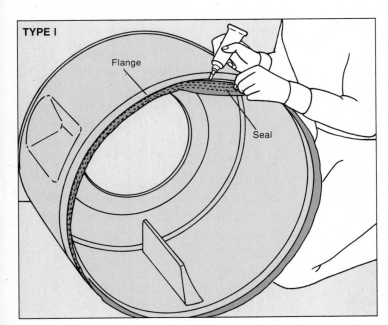

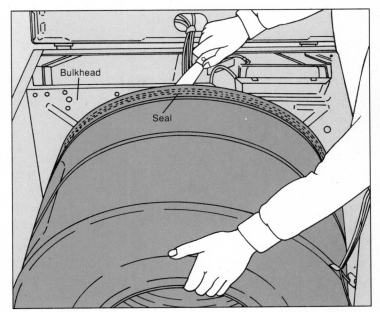

5 **Replacing the rear drum seal (Type I dryers).** With the drum removed, check the felt-and-plastic seal around its back edge; if the seal is damaged, scrape it off the drum with a putty knife. Clean any adhesive off the drum flange with a rag and paint thinner (do not use lacquer thinner). Slip a new seal around the drum flange, its stitched edge in. Lift the inner edge of the seal and apply a bead of adhesive around the drum flange, as shown, pressing the seal down as you go. Let the adhesive set for one hour, then reinstall the drum.

6 **Reinstalling the drum.** For a Type I dryer, slide the drum in through the front of the dryer, and rest the rear drum flange on the support rollers. Rethread the drive belt *(page 121)*. Seat the rear seal against the bulkhead by inserting a putty knife between the seal and the bulkhead, as shown; rotate the drum a full revolution to be sure the seal edge is not pinched. Replace the dryer panels.

On a Type II dryer, seat the front groove of the drum on the slides and, from the back, snap on the drum shaft retainer. Reinstall the grounding strap and the drive belt, and replace the dryer panels.

TESTING AND REPLACING THE HEATER ELEMENT (Type I dryers)

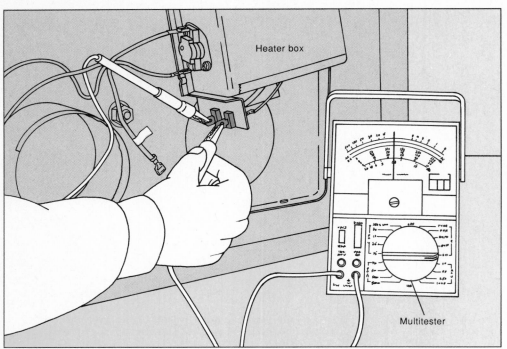

Heater box

Multitester

1 **Testing the heater element.** Unplug the dryer and remove the rear panel *(page 118)*. Disconnect the wires to the heater terminals and label their positions with masking tape. Set a multitester to the RX1 scale. If the heater has two terminals, touch one probe to each terminal, as shown; the meter should show 5 to 50 ohms. Then touch one probe to the heater box and the other to each terminal in turn; the needle should not move. If the heater has three terminals, touch one probe to the middle terminal and the other probe to the outer terminals in turn; the meter should show 10 to 40 ohms. Then, to test for a ground, touch one probe to the heater box and the other to each terminal in turn; the meter should not move. If the element fails any of these tests, proceed to step 2 to remove and replace it.

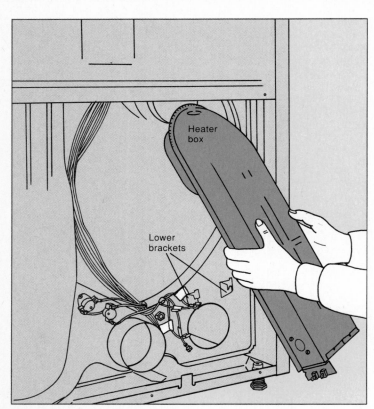

Heater box

Lower brackets

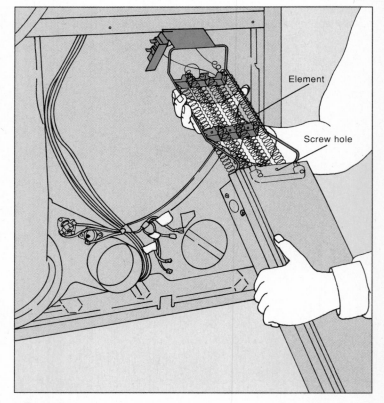

Element

Screw hole

2 **Removing the heater box.** Raise the dryer top *(page 117)* and unscrew the bracket holding the heater box to the bulkhead. Unscrew the thermostat from the side of the heater box. Lift the heater box slightly to free it from the lower brackets, and pull it down and out from the rear of the dryer, as shown.

3 **Installing a new heater element.** Remove the screw holding the element in the heater box and carefully pull out the element. Slide a new element into the box, same side up. Be sure the coils do not rub against the sides of the box. Insert and tighten the screw. Slide the heater box up into the rear of the dryer, hook the slots onto the lower brackets, and reattach the upper bracket. Screw the thermostat back on the heater box and replace the rear panel.

TESTING AND REPLACING THE HEATER COILS (Type II dryers)

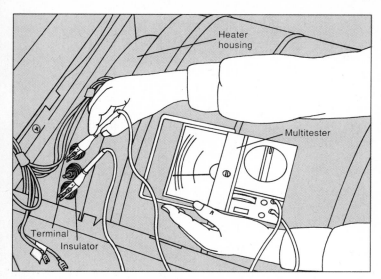

1 **Testing the heater coils.** Unplug the dryer and raise the dryer top *(page 117)*. Disconnect the wires from the insulator terminals on the heater housing and label their positions with masking tape. Set a multitester to the RX1 scale. Touch one probe to the left (common) terminal, and the other probe to each of the other terminals in turn. In each case the multitester needle should move, showing continuity. Next, to test for a ground, touch one probe to the heater housing and the other probe to each terminal in turn. The needle should not move. If either coil fails any of these tests, replace *both* coils.

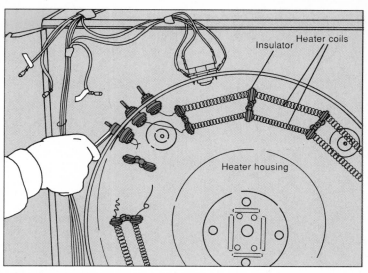

2 **Removing the old heater coils.** Take out the drum *(page 122)*. Check the heater coils for breaks, burns or broken insulators. To remove the coils, use wire cutters to snip their ends near the terminals, and carefully unthread the coils through the insulators. Unscrew the nuts and washers holding the terminals to the heater housing, as shown; discard the old terminals, but keep the ceramic insulators.

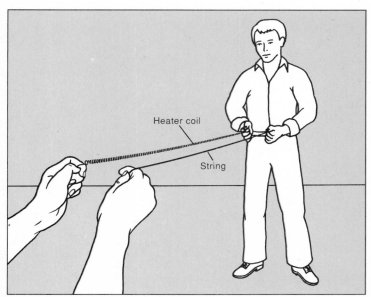

3 **Stretching a replacement coil.** New heater coils must be stretched to the required length before installation. The heater coil replacement kit indicates the length needed for your dryer model. Most often, the outer coil is stretched 63 inches to produce a relaxed length of 40 inches; the inner coil is stretched 56 inches, producing a 35-inch coil. To stretch a coil, cut a piece of string to the correct length, and hold one end of the coil and the string in each hand. Pull the coil slowly and evenly until the string is taut (or have another person help you, as shown). Do not stretch the coil in sections; the resulting unevenness causes hot spots. Cut and bend the ends of the outer coil to form 1-inch hooks, and bend the ends of the inner coil to form 1/2-inch hooks.

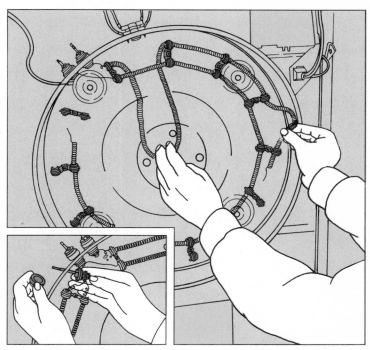

4 **Installing the heater coils and terminals.** Slip a nut and a washer on one end of a new terminal, hook the end of the heater coil around the terminal, and sandwich it tightly with another washer and nut. Place a ceramic insulator on the terminal and insert the terminal through a hole in the heater housing: The inner coil goes to the middle hole and the outer coil goes to the right-hand hole. Place a second insulator on each terminal from outside the housing and secure it with a nut. Gently thread each new coil clockwise through the insulators, as shown. Wrap the two free ends around a single terminal, secure them with washers and nuts, and install the terminal with insulators in the remaining hole *(inset)*.

SERVICING THE GAS BURNER (Gas dryers)

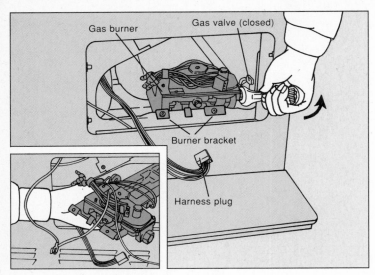

Gas burner · Gas valve (closed) · Burner bracket · Harness plug

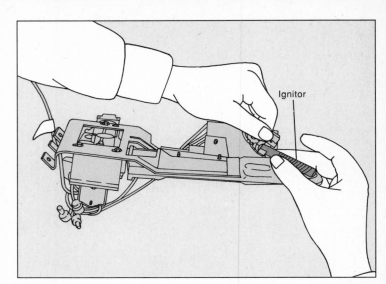

Ignitor

1 Removing the gas burner. Unplug the dryer and remove the toe panel *(page 118)* or pry open the small gas burner access panel on the front of the dryer. Shut off both the main gas valve to the dryer and the gas valve leading to the burner. (A gas valve is closed when the handle is perpendicular to the pipe.) Disconnect the multi-wire harness plug and any other wires leading to the gas burner. Using an open-end wrench, unscrew the union nut between the burner and the gas pipe by pushing the wrench away from you, toward the dryer *(above)*. The end of the gas pipe will separate from the burner; lay the pipe aside. Remove the two hex-head screws holding the burner bracket to the dryer cabinet. Carefully lift out the burner *(inset)*.

2 Testing and replacing the ignitor. Turn the burner over and carefully use your thumb to spread the mounting clips that grip the ignitor *(above)*. The ignitor is extremely brittle; disengage it with great care. On some burners the ignitor is screwed in place; use a nut driver to release it. Set a multitester on the RX100 scale. Touch a probe to each side of the silver end of the ignitor, between the fins. The meter should show 50 to 400 ohms; if not, replace the ignitor. Carefully spread the clips and gently insert the new ignitor, or screw the ignitor in place.

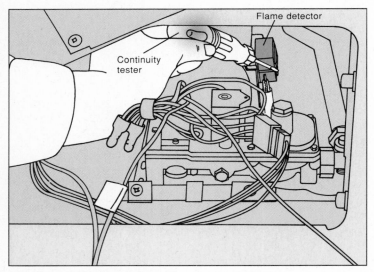

Flame detector · Continuity tester

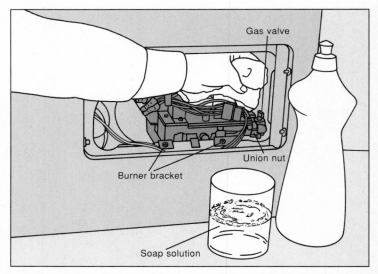

Gas valve · Union nut · Burner bracket · Soap solution

3 Testing and replacing the flame detector. The flame detector is usually screwed to the side of the combustion chamber, behind the gas burner, but some are located on the burner itself. Disconnect the wires to the detector. Touch one probe of a continuity tester to each terminal of the flame detector; the tester should light, showing continuity; if not, replace it. If you can reach the flame detector with a nut driver, unscrew it and install a new one. In most cases, however, you must first remove the gas burner and the funnel-shaped combustion chamber to reach the flame detector. If both the ignitor and flame detector test OK, take the gas burner assembly to a service center for replacement or repair, or call for service.

4 Reinstalling and testing the gas burner. Screw the burner bracket to the dryer cabinet and reconnect all wires. Reattach the gas pipe to the burner by hand, turning the union nut counter-clockwise to the burner, then tighten with a wrench. To test for a gas leak, prepare a half-and-half solution of dishwashing liquid and water. Turn the gas valve handle to the ON position (parallel to the gas pipe), then turn on the gas supply to the dryer. Brush the soap-and-water solution on all joints. If bubbles form, shut off the gas, loosen and retighten the joint, turn on the gas and test again. If bubbles still form, turn off the gas to the dryer and call for service.

SERVICING THE MOTOR AND BLOWER (Type I dryers)

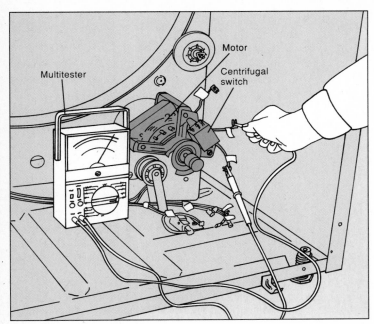

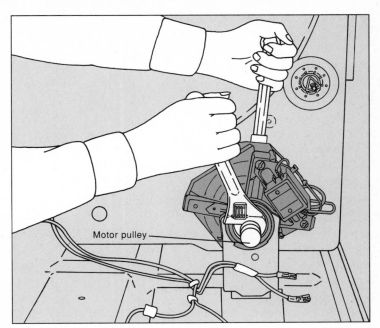

1 **Testing the motor.** Unplug the dryer. Raise the top *(page 117)* and remove the toe panel, if any, and the front panel *(page 118)*. Remove the drive belt and idler *(page 121)* and take out the drum *(page 122)*. You now have access to the motor and blower. Disconnect the wires to the motor and centrifugal switch, and label their positions with masking tape for reassembly. To test the motor, set a multitester to the RX1 scale. Connect one probe to the yellow wire leading to the motor, and the other to the blue wire. The meter should indicate 1 to 5 ohms of resistance. Remove the probe from the yellow wire and attach it to the black wire. The meter should again indicate 1 to 5 ohms of resistance. If the motor fails either test, remove it and take it to a service center, or call for service.

2 **Releasing the blower wheel from the motor shaft.** Using two adjustable wrenches, grip the motor shaft in front of the motor (behind the motor pulley) and the blower-wheel hub at the back of the motor. Hold the blower wheel stationary as you turn the motor shaft clockwise toward the side of the dryer, until the blower wheel is free of the shaft.

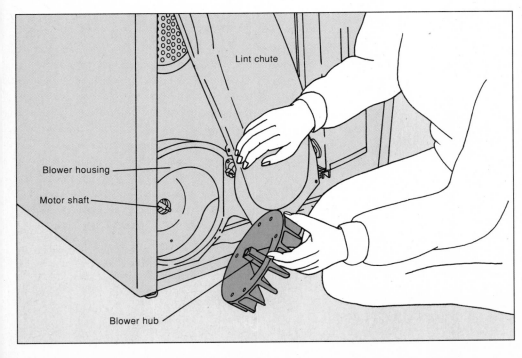

3 **Inspecting and replacing the blower.** Remove the rear panel *(page 118)* to reach the lint chute and blower housing. Use a nut driver to remove the four screws holding the lint chute to the blower housing; if necessary, also remove two screws at the upper corner of the bulkhead. Push the lint chute up and over toward the middle of the dryer, exposing the blower. Remove the blower from the housing and check for lint and foreign objects. Look for damage to the threads inside the blower hub and to the blower fins, and install a new blower if necessary. If you plan to remove the motor, go to step 4. If you are replacing only the blower, loosely thread the hub of the new blower onto the motor shaft. From inside the dryer, tighten the shaft using the same double-wrench technique as in step 2. Screw the lint chute in place and replace the rear panel. Reconnect the wires to the motor and the centrifugal switch. Install the drum, idler and belt, and replace the dryer panels.

SERVICING THE MOTOR AND BLOWER (Type I dryers, continued)

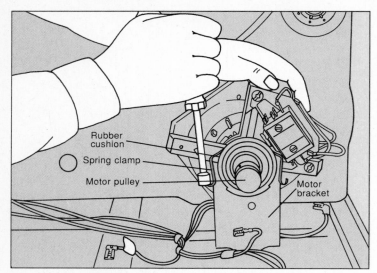

Rubber
cushion

Spring clamp

Motor pulley

Motor
bracket

4 **Releasing the motor clamps.** Unscrew the green ground wire from the motor and label it with tape. Straplike spring clamps secure the round rubber cushions at the front and back of the motor to the motor bracket. To release each clamp, press down on the hooked end with a nut driver and snap the clamp off the motor bracket *(above)*. The motor can now be lifted out for professional servicing or replacement. If you are installing a new motor, remove the motor pulley from the shaft by loosening the set screw with a hex wrench, and reinstall the pulley on the shaft of the new motor.

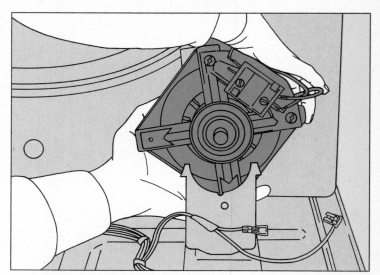

5 **Reinstalling the motor.** Position the motor with the threaded end of the shaft to the rear of the dryer *(above)*. Set the rubber cushions in the motor brackets, fitting the tab on the front cushion into the slot in the front bracket. Place the clamps across the rubber cushions and snap them onto the brackets with a nut driver. Thread the blower onto the rear motor shaft as in step 3. Reconnect the wires to the motor and the centrifugal switch. Install the drum, idler and belt, and replace the dryer panels.

SERVICING THE MOTOR AND BLOWER (Type II dryers)

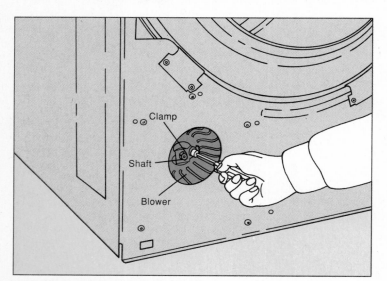

Clamp

Shaft

Blower

1 **Adjusting the blower.** Remove the front panel *(page 118)* to expose the blower opening at the lower left corner of the cabinet. Clean any obstructions from the blower and housing. Turn the wheel by hand; if it wobbles, binds or rubs against the housing, reseat it on the shaft. Loosen, but do not remove, the hex-head screw that secures the blower clamp *(above)*. Reposition the blower on the shaft and tighten the clamp. If the problem is not solved, remove the motor-and-blower assembly *(step 3)*.

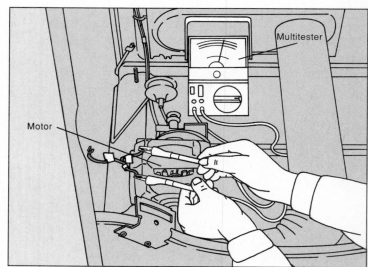

Multitester

Motor

2 **Testing the motor.** Remove the drum *(page 122)*. Set a multitester to the RX1 scale. Connect one probe to the motor's orange wire, and the other to the blue wire. The meter should indicate 1 to 5 ohms of resistance. Remove the probe from the orange wire and attach it to the black wire. The meter should again indicate 1 to 5 ohms of resistance. If the motor fails either test, remove it *(step 3)* and take it to a service center, or call for service.

SERVICING THE MOTOR AND BLOWER (Type II dryers, continued)

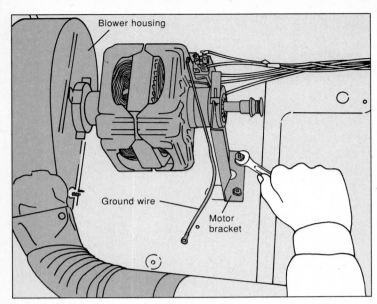

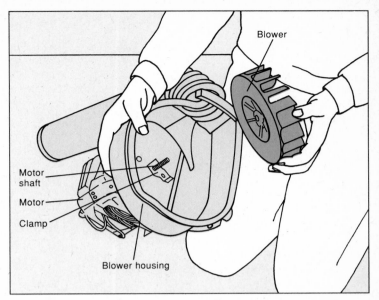

3 **Removing the motor-and-blower assembly.** Use a nut driver to remove the screws around the blower opening on the front of the dryer cabinet. Disconnect the green ground wire from the motor and unscrew the two nuts holding the motor bracket to the dryer floor *(above)*. Lift the motor-and-blower assembly to expose the blower wheel. If you plan to remove the motor, disconnect all wires to the motor and centrifugal switch, and label their positions with masking tape. Remove the assembly from the dryer. Before taking the motor for repair or replacement, also remove the blower, blower housing and motor bracket.

4 **Removing and replacing the blower.** Remove the mounting screw on the center clamp of the blower, and pull the clamp and blower off the motor. The clamp has a back half behind the blower; note its position and remove it from the shaft. To replace the blower wheel, place the back half of the clamp on the shaft in the proper position, and hold the front half of the clamp in the same position on the front of the new blower. Place the blower on the shaft, aligning the two halves of the clamp, and screw the clamp together loosely. Reinstall the motor-and-blower assembly, screwing the motor bracket to the dryer floor and the blower housing to the dryer cabinet. Adjust the blower and tighten the clamp as in step 1.

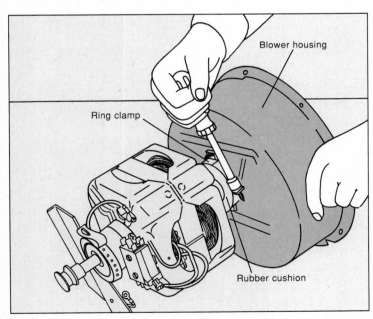

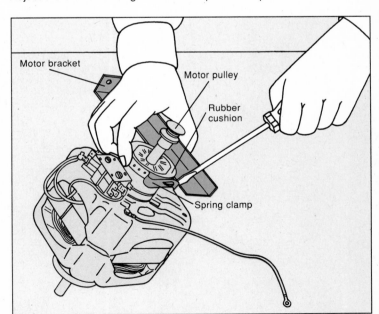

5 **Freeing the blower housing from the motor.** The blower housing is secured to the motor by a ring clamp that grips a rubber cushion. With the blower removed, use a nut driver to loosen the two screws on the clamp *(above)*, and pull the blower housing off the motor.

6 **Removing the motor bracket.** Using a screwdriver, pry off the spring clamp that holds the motor bracket to the rubber cushion on the motor, as shown. Remove the motor pulley from the motor shaft by loosening the set screw with a hex wrench. The motor is now ready to be replaced or serviced professionally. To reassemble the dryer, screw the motor pulley to the shaft and clamp on the motor bracket and blower housing. Attach the blower and reinstall the motor in the dryer. Reconnect the wires and replace the dryer panels.

TOOLS & TECHNIQUES

This section introduces basic tests and repairs that are common to almost all major appliances, from checking power cords to patching minor nicks and scratches. Illustrations reveal the electrical, plumbing and gas hookups that allow various appliances to function, and demonstrate how to work safely with each system. Wiring diagrams and timer charts—the invaluable road maps to an appliance's wiring—are explained on page 138. Advice on buying replacement parts, and how to deal with a professional when all else fails, appears on page 141.

You can handle most appliance repairs with the basic tool kit shown below. For the best results, buy the best tools you can afford, use the right tool for the job, and take the time to care for and store them properly. To prevent rust, clean metal tools after every use with a rag moistened with a few drops of light machine oil. (Don't oil handles; you could lose your grip.) Rust is best removed by buffing with fine steel wool and oil or kerosene. Avoid electrical shock by using only pliers and screwdrivers with insulated handles, or handles wrapped in electrical tape. Protect tools in a sturdy plastic or metal toolbox, with a secure lock if stored around children. Specialized tools, such as a multitester, can often be rented from the same appliance-parts supply stores where you will buy replacement parts.

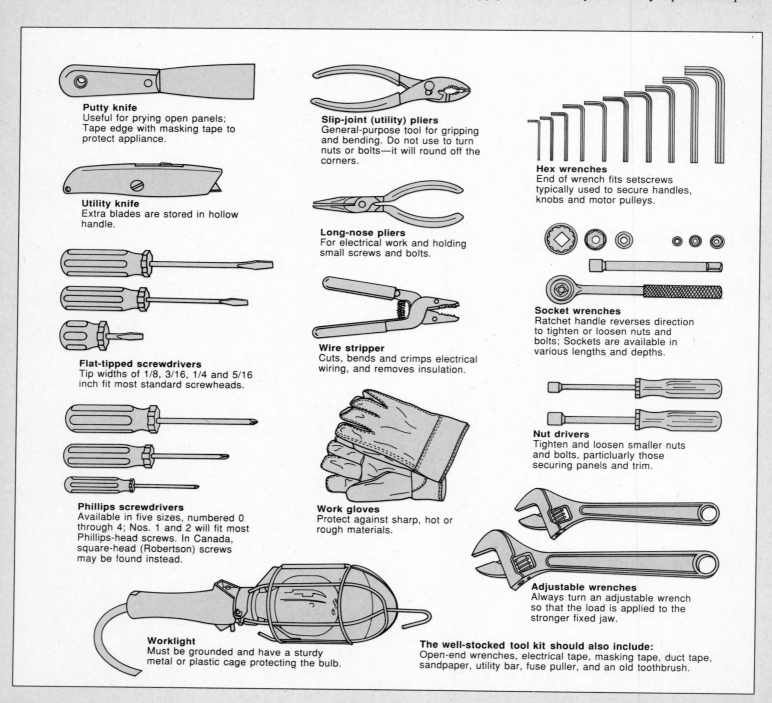

Putty knife
Useful for prying open panels; Tape edge with masking tape to protect appliance.

Utility knife
Extra blades are stored in hollow handle.

Flat-tipped screwdrivers
Tip widths of 1/8, 3/16, 1/4 and 5/16 inch fit most standard screwheads.

Phillips screwdrivers
Available in five sizes, numbered 0 through 4; Nos. 1 and 2 will fit most Phillips-head screws. In Canada, square-head (Robertson) screws may be found instead.

Worklight
Must be grounded and have a sturdy metal or plastic cage protecting the bulb.

Slip-joint (utility) pliers
General-purpose tool for gripping and bending. Do not use to turn nuts or bolts—it will round off the corners.

Long-nose pliers
For electrical work and holding small screws and bolts.

Wire stripper
Cuts, bends and crimps electrical wiring, and removes insulation.

Work gloves
Protect against sharp, hot or rough materials.

Hex wrenches
End of wrench fits setscrews typically used to secure handles, knobs and motor pulleys.

Socket wrenches
Ratchet handle reverses direction to tighten or loosen nuts and bolts; Sockets are available in various lengths and depths.

Nut drivers
Tighten and loosen smaller nuts and bolts, particularly those securing panels and trim.

Adjustable wrenches
Always turn an adjustable wrench so that the load is applied to the stronger fixed jaw.

The well-stocked tool kit should also include:
Open-end wrenches, electrical tape, masking tape, duct tape, sandpaper, utility bar, fuse puller, and an old toothbrush.

DIAGNOSING ELECTRICAL PROBLEMS

Troubleshooting with electrical testers. From its source at the wall outlet, an electrical current travels through the power cord, switches, timers, thermostats, heating elements, motors and other parts that make up the electrical system of a major appliance. A part may allow the current to pass through it unimpeded (a power cord, a switch in the ON position); it may interrupt the current completely (a switch in the OFF position, a blown fuse); or it may partially block the current (to heat a range element or propel a motor).

Throughout this book two simple testing devices—the continuity tester and the multitester—are used to determine whether an electrical part is doing its job. Both testers are battery-powered, and send a small electrical current through the part being tested.

The continuity tester simply indicates whether all the current is passing through the part *(continuity)* or whether the current is partially or completely blocked *(resistance)*. A multitester, also called a volt-ohmmeter, provides a precise measurement of the amount of resistance encountered by the current.

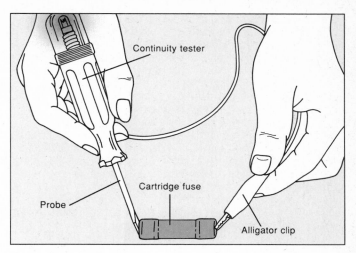

Using a continuity tester. First check the tester's battery by touching the alligator clip to the probe; the bulb should light. To test for an uninterrupted circuit (through a cartridge fuse, in this example), place the clip against one end of the fuse and the probe against the other end. If current from the tester meets any resistance, the tester will not light; if the circuit is complete, the bulb will glow.

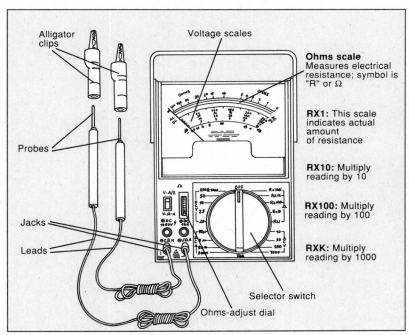

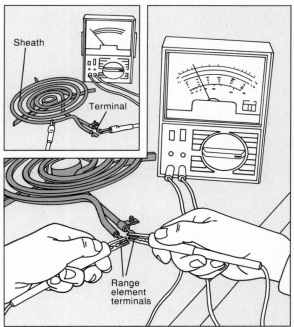

Using a multitester. A multitester gives an exact reading of the amount of resistance present in an electrical circuit, as measured in ohms. Zero ohms indicates continuity—the complete lack of resistance. For low resistance readings, up to 200 or 500 ohms (depending on your tester), set the selector switch of the multitester to RX1. When greater resistance is being tested, change the setting to RX10 (and multiply the reading by 10), RX100 (multiply by 100) or RXK (multiply by 1,000).

To use the multitester, first set the selector switch on the specified scale. To ensure an accurate reading, "zero" the multitester by clamping the probes together. The needle will sweep from left to right down to ZERO; turn the ohms-adjust dial until the needle aligns directly over ZERO. In this example, an electric range heating element is being tested for resistance. Set the multitester to the RX1 scale and place the probes against each element terminal *(above, right)*. A properly working range element will cause the needle to sweep upscale, indicating partial resistance. If not, the element is faulty and should be replaced. To test the element for grounding, set the multitester on the RXK scale. Touch one probe to a terminal and the other to the coil sheathing. *(inset)*. If the needle moves at all, there is a potentially dangerous current leakage and the element should be replaced.

WORKING WITH ELECTRICITY

From the utility company's power lines, electricity enters your home through an electric meter and into the main service panel—a circuit-breaker panel in newly wired houses and a fuse panel in older wiring systems. From the service panel, electricity is distributed throughout the house by a number of separate circuits. A circuit is the closed path that electrical current follows from a power source (service panels and outlets) through various switches, fixtures and appliances, and then back to the source.

The strength, or pressure, of the electrical current moving through a circuit is measured in volts. The rate of current flow is measured in amperes (amps). Low-capacity circuits of 120 volts and 15 amperes are adequate for most lamps, TV sets and small appliances. Most major appliances require heavier-capacity circuits of 120 volts and 20 amperes. Individual circuits of 240 volts and 30 to 50 amperes are required for clothes dryers and electric ranges. These voltages are nominal; the actual voltage delivered by the power company may vary by 10 percent.

A third measurement—wattage—is calculated by multiplying volts by amperes. Wattage describes how much electricity is being converted by an appliance into another form of energy such as heat or motion. Thus a 24-amp clothes dryer on a 240-volt circuit consumes about 5,600 watts of energy. Total energy consumption is calculated by multiplying the power (expressed in 1,000-watt units called kilowatts) by the amount of time an appliance is kept running. The result is kilowatt-hours (kwh). If the 5600-watt clothes dryer is used for one hour, it will consume 5.6 kilowatt-hours, which will be tallied by the meter and added to your electric bill.

Circuit breakers and fuses protect the individual electrical circuits in your home. Throughout this book, you will be directed to shut off power to an appliance either by pulling its plug, tripping its circuit breaker or removing its fuse (below). **Caution:** When working on the service panel, always keep one hand free—or stand on a dry wooden board or rubber mat—to avoid the possibility of shock due to accidental grounding.

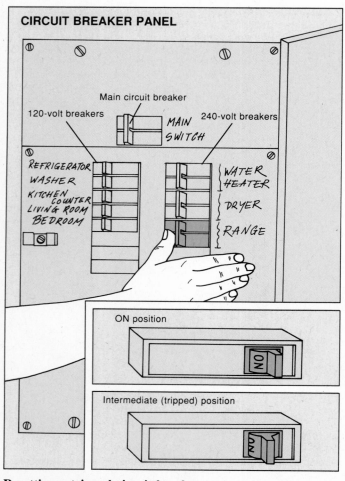

CIRCUIT BREAKER PANEL

Resetting a tripped circuit breaker. When a circuit is overloaded, the circuit breaker toggle automatically flips to OFF or an intermediate position, shutting off power. When this happens, first locate and correct the problem in the circuit or appliance. Then reset the circuit breaker by moving it fully to the OFF position, then back to ON.

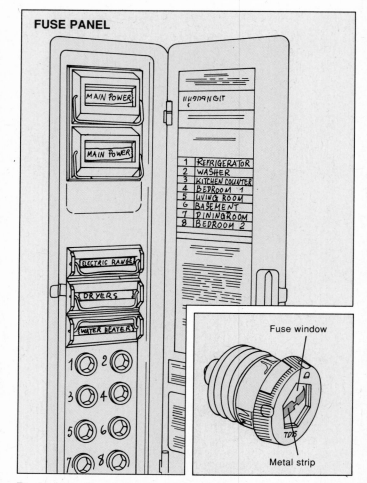

FUSE PANEL

Replacing a fuse. When a circuit protected by a plug-type fuse overloads, the metal strip inside the fuse melts. If there is a short in the circuit, the fuse window will become blackened. Remove the fuse by turning it counterclockwise; replace it with a fuse of identical amperage. Cartridge fuses, which protect circuits up to 240 volts, show no signs of damage and must be checked with a continuity tester (page 131).

POWER CORDS AND TERMINAL BLOCKS

Carrying current from outlet to appliance. Perhaps the most common appliance problem is that electricity cannot get to the machine. If an appliance does not work at all, first check the main service panel for a blown fuse or tripped circuit breaker *(page 132)*. If the problem is not found there, next inspect the power cord and plug *(below)*.

Power cords are made of separately insulated wires, sheathed together, that carry electricity from the wall outlet through a plug to the appliance and back. Usually a simple visual inspection will locate any faults. A broken or wobbling plug prong, cracked or burned insulation, or a loose connection at the cord sleeve all indicate a break in the circuit—and that both the cord and plug should be replaced. Never splice a power cord.

Invisible damage, such as a break or short in the wiring, is harder to trace. Shaking the cord sometimes briefly reestablishes a lost connection and may point to trouble in the cord or the terminal block. A more reliable method is to test the cord with a continuity tester or a multitester.

Since the early 1970s, all major appliances have been manufactured with grounded plugs in accordance with national and local electrical codes. Power cords should always be plugged into matching three- or four-prong outlets. Grounding adapter plugs should never be used to adapt an appliance to a two-prong outlet—the plug is useless unless the outlet itself is grounded. If you suspect that the outlet intended for an appliance is not grounded, have an electrician test it and, if necessary, install a grounded outlet or a ground-fault circuit interrupter.

To inspect and test a power cord, pull the appliance away from the wall, taking care not to disturb supply pipes or exhaust ducts. Gas ranges and dryers with flexible supply hoses can be pulled away from the wall, but do not move a gas appliance with a rigid supply pipe—call the gas company or a service technician.

SERVICING 120-VOLT POWER CORDS

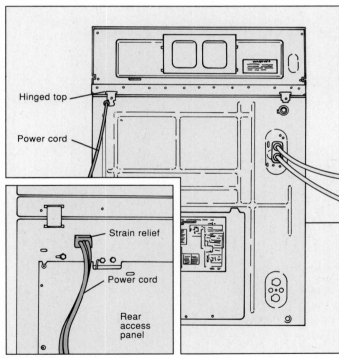

Hinged top

Power cord

Strain relief

Power cord

Rear access panel

1 **Access to the power cord connector.** The power cord enters a 120-volt appliance through the back of the machine, and can be accessed in one of two ways: By lifting the top *(above)* or by removing a cover plate or rear access panel *(inset)*. Unplug the machine; remove the nuts or screws securing the cover plate or rear access panel, or open the top of the appliance. If you don't see the power cord connection, trace the cord into the machine to locate it. (See also the access steps for each appliance.)

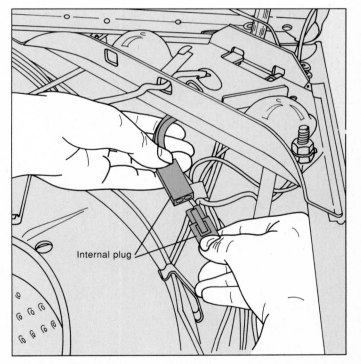

Internal plug

2 **Disconnecting the power cord.** Power cords for most 120-volt appliances (except refrigerators) are connected to the internal wiring of the machine by a small plug *(above)*. Pull apart the two halves of the plug and inspect the cord terminals for damage. If the terminals are corroded or out of alignment, or the plug itself is cracked or burned, power cannot reach the appliance properly and the cord should be replaced *(step 4)*.

SERVICING 120-VOLT POWER CORDS (continued)

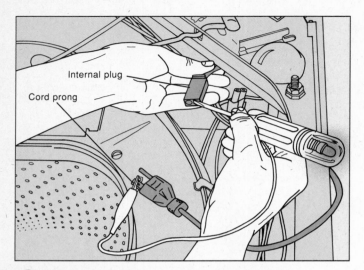

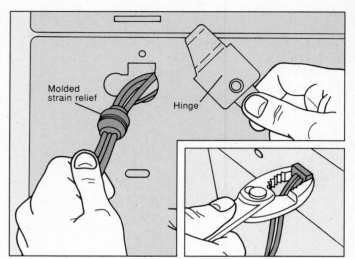

3 **Testing the power cord.** Attach the clip of a continuity tester to one of the flat prongs on the power cord. Touch the tester probe to one of the terminals in the internal plug, then to the other terminal. The tester should glow against one, but not both terminals. Repeat this test with the other flat prong and replace the cord if faulty. Next, detach the ground wire connected to the cabinet, and test it for continuity against the round prong. If the tester does not light, replace the cord.

4 **Replacing the power cord.** Some power cords have a molded strain relief that prevents the cord from being pulled from the machine and damaging the wiring. To free the cord, remove the hinge that holds the cord to the cabinet *(above)*. Connect the new cord to the internal wiring, attach the ground wire to the cabinet and position the strain relief on the back panel before retightening the hinge. If the strain relief is attached to the cabinet, squeeze it with pliers to free the cord *(inset)*.

SERVICING 240-VOLT POWER CORDS

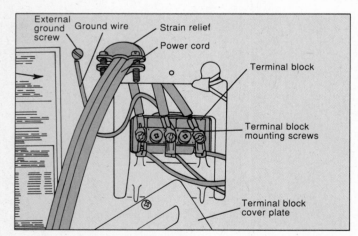

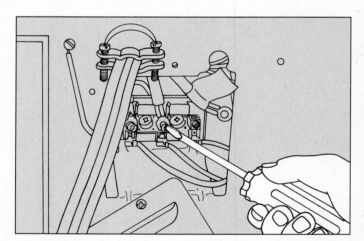

1 **Access to the terminal block.** The power cords of 240-volt appliances are connected to the machine's internal wiring via a terminal block. (Some 120-volt refrigerators also have terminal blocks.) If the appliance fails, or if a heating element doesn't heat, both the power cord and terminal block should be checked for damage. Turn off power to the appliance and remove the cover plate where the cord enters the back of the machine. Inspect the terminal block for loose, burned, broken or corroded wires. At any visible sign of damage, replace the terminal block.

2 **Disconnecting the power cord.** To test the power cord for internal damage, first disconnect it from the terminal block. Label the wires for reassembly and, depending on your machine, either pull off the wire connectors or remove the screws or hex nuts holding the wires in place, as shown.

SERVICING 240-VOLT POWER CORDS (continued)

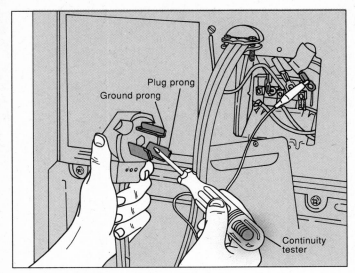

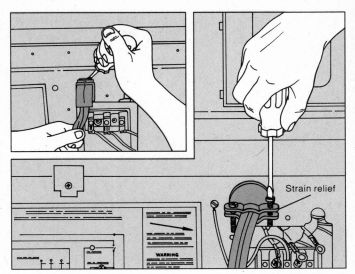

3 **Testing the power cord.** Touch the probe of a continuity tester to one of the flat prongs on the power cord plug. Touch the other tester probe to the terminals of the three wires in turn. The tester should glow against only one of the terminals. Repeat this test with the other flat prong; the tester should glow against a different terminal. Replace the cord if faulty.

Next, touch the tester probe to the ground prong, and the other probe to the terminal of the middle wire; the tester should light. If you have a four-prong, four-wire power cord, test the top middle ground plug against the ground wire. Test the bottom middle plug against the middle wire to the terminal block. The tester bulb should glow; if not, replace the cord.

4 **Replacing the power cord.** To replace a power cord with a four-prong plug, first remove the screw securing the ground wire to the cabinet. If the cord is protected by a metal strain relief, loosen it *(above)* and pull the cord through the back of the cabinet. Newer appliances may have a plastic strain relief that is molded on the cord; pry the cord free with a flat-tipped screwdriver *(inset)*. If you are replacing the terminal block, go to step 5. Otherwise, feed the new power cord through the back of the cabinet, wire it to the terminal block, and replace the cover plate.

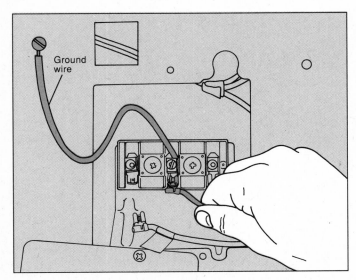

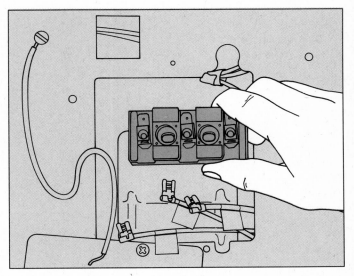

5 **Disconnecting the terminal block.** If the terminal block appears cracked, corroded or burned, replace it. The machine's internal wiring is attached to the block either with push-on connectors, as shown, or with screw-and-eyelet connectors. Label the wires for reassembly, then pull off or unscrew the connectors. Disconnect the ground wire or metal grounding strap leading from the terminal block to the cabinet.

6 **Replacing the terminal block.** Remove the terminal block mounting screws or nuts and lift the block from the cabinet *(above)*. Install a replacement terminal block with the proper rating for your make and model of appliance. Reattach the ground wire or strap, connect the internal wires and power cord wires to their proper terminals and replace the cover plate.

REPAIRING DAMAGED WIRING

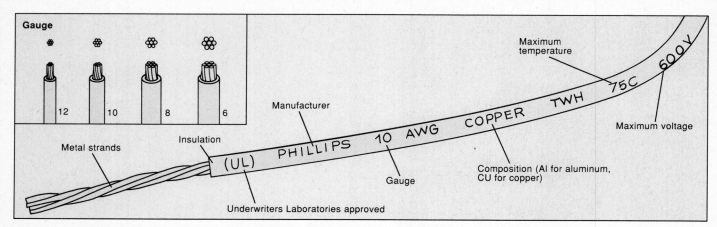

Gauge

12   10   8   6

Metal strands

Insulation

(UL)   PHILLIPS   10 AWG   COPPER   TWH   75C   600V

Manufacturer

Underwriters Laboratories approved

Gauge

Composition (Al for aluminum, CU for copper)

Maximum temperature

Maximum voltage

Reading a wire. The internal wiring of a major appliance *(above)* is made of many thin copper or aluminum strands wrapped together and sheathed in thermoplastic or fiberglass insulation (Older appliances may have asbestos insulation., The diameter, or thickness, of the wire is indicated by a gauge number, usually printed on the insulation. The smaller the number, the thicker the wire and the more current it can carry. Most appliances use No. 6 to No. 12 wires *(inset)*. The particular gauge is determined by the wire's function and the amount of current that it must carry. Electric dryers and ranges, which run on 240 volts, require lower-

gauge wiring than 120-volt appliances such as washers or refrigerators. If a wire is burned, broken, corroded or shows resistance when checked with a multitester or continuity tester, it should be repaired or replaced. A damaged section of wire may be cut off, and a new wire spliced on. A permanently wired part that is faulty must be cut off, and a new part spliced in its place. In either case, the wires can be joined in one of two ways: by twisting them into a wire cap, or by inserting them into a crimp connector. Before working with wiring, shut off power to the appliance.

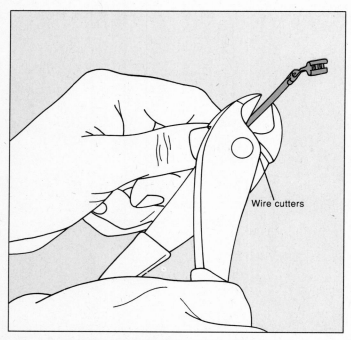

Wire cutters

1 Removing damaged wiring. Most damage, particularly burned or oxidized terminals, is visible. But if you suspect a wire has hidden damage, run your fingers along it from terminal to terminal while gently bending and twisting it. If you locate a bump or sudden limpness, that section should be snipped off with wire cutters *(above)*.

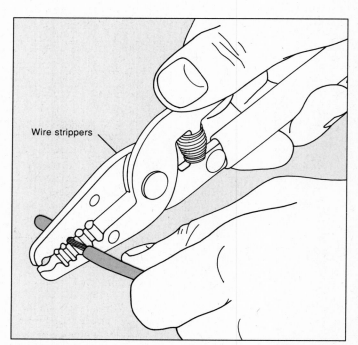

Wire strippers

2 Stripping insulation. Insert the wire into a matching slot on a pair of wire strippers. (The gauges of the wire and slot must be the same.) Close the tool and twist it back and forth until the insulation is severed and can be pulled off the wire *(above)*. Strip about 1/4 inch of insulation from each wire to be spliced. If you are splicing two different wires, proceed to step 3; if you are rejoining the same wire, or replacing a terminal connector, go to step 5.

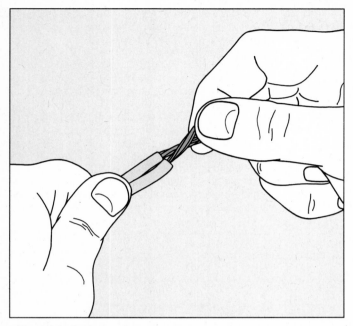

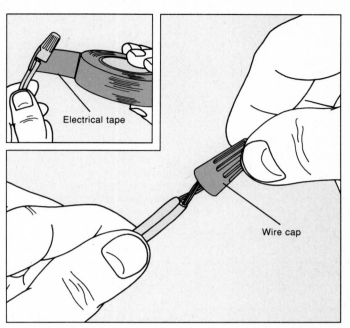

Electrical tape

Wire cap

3 **Twisting the wires together.** Hold the wires together, as shown, and twist them clockwise into a "pigtail." Higher-gauge (thin) wire can be twisted by hand. To twist lower-gauge (thick) wire, long-nose pliers may be needed.

4 **Securing the splice.** Slip a wire cap over the pigtail and screw the cap clockwise until it is tight and no bare wire remains exposed *(above)*. To make sure that the wires will not jar loose, secure the cap with electrical tape *(inset)*. Wrap the tape around the base of the cap, then once or twice around the wires, and finally around the base of the cap again.

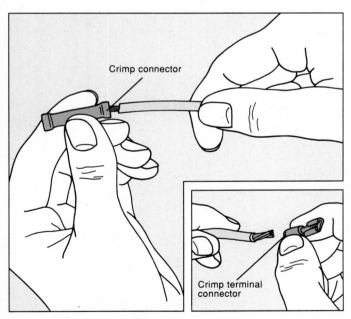

Crimp connector

Crimp terminal connector

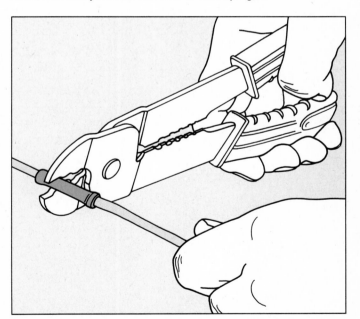

5 **Splicing wires with crimp connectors.** Be sure to use a crimp connector of the same gauge as the broken wire. (Crimp connectors are color-coded by thickness.) If you are joining wires, cut and strip the two wire ends *(step 2)*, and insert them into the connector, as shown. To replace a terminal connector, choose a crimp-style connector *(inset)* of the appropriate gauge and type.

6 **Securing the connection.** Use the crimping jaws of a wire stripper to pinch the connector near both ends. Check that no bare wires are visible, and gently tug on the wires to make sure the connections are secure, before pushing the wiring back into the appliance.

READING WIRING DIAGRAMS AND TIMER CHARTS

Though the wires inside a major appliance may resemble a confusing maze, there are several aids to deciphering the machine's internal workings. One is the wiring diagram, a schematic drawing that uses various symbols to show how the machine's wires and electrical components are connected. Another is the timer chart (or cam chart), showing which switches are in operation during an appliance's various cycles, and for how long. On most large appliances, both diagrams will be found glued to an access panel or tucked inside the control console. If yours are missing, write or call the manufacturer for copies.

The wiring diagram is a road map of the circuit, or loop, by which electrical current travels from the power cord, through a sequence of switches, timers, motors or heating elements, and back to the power source. In some cases the lines, indicating wires, split off in more than one direction, depicting separate loops within the main circuit. By tracing a circuit from start to finish, you will be able to test the components in their correct sequence, and thus pinpoint your problem quickly. While the styles of wiring diagrams vary, a number of symbols are common to all of them (below).

Say, for example, that you close the door of your electric dryer, set the timer and press the start button, but the machine does not work. First, make sure the dryer is plugged in and receiving power. Then consult the wiring diagram to begin testing its electrical components in order.

The diagram on page 139 is of a typical dryer. Start at the power source, shown at the top by the symbol for a grounded three-prong plug. Since dryers require a large current, there are three incoming wires, here marked "L1" for Line 1, "L2" for Line 2, and "N" for the neutral wire. The 240-volt current that flows between L1 and L2 powers the heating element of the dryer. Since the problem is not a lack of heat, concentrate on the 120-volt circuit (highlighted in the illustration) that powers the appliance's motor and some of the switches.

WIRING DIAGRAM SYMBOLS AND TIMER CHART

COMPONENT	SYMBOL
MOTOR (SINGLE-SPEED)	
MOTOR (MULTI-SPEED)	
STARTING CAPACITOR	
WIRE TERMINAL	
PERMANENT WIRE CONNECTION	
WIRES CROSS OVER (NO CONNECTION)	
WIRES CROSS OVER (NO CONNECTION)	
HEATER (WATTAGE SHOWN)	2800w
HEATER (WATTAGE SHOWN)	5600w
THERMOSTAT	
THERMOSTAT	
OVERLOAD PROTECTOR	
CENTRIFUGAL SWITCH (OR MOTOR START SWICHT)	
TIMER SWITCH	

COMPONENT	SYMBOL
AUTOMATIC SWITCH OR MANUAL SWITCH	
PUSH-TO-START SWITCH	
RESISTOR (OHMS SHOWN)	5000Ω
TERMINAL BOARD	
BUZZER	
RELAY	

COMPONENT	SYMBOL
NEON LIGHT	
FLUORESCENT LIGHT	
INCANDESCENT LIGHT	
3-PRONG PLUG	
GROUND (EARTH)	
CHASSIS	

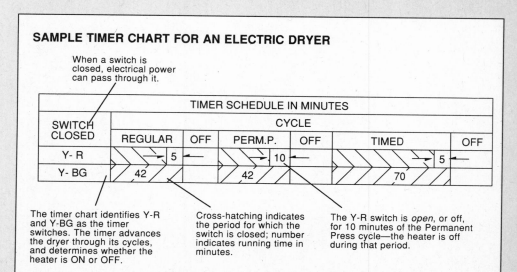

SAMPLE TIMER CHART FOR AN ELECTRIC DRYER

When a switch is closed, electrical power can pass through it.

SWITCH CLOSED	TIMER SCHEDULE IN MINUTES					
	CYCLE					
	REGULAR	OFF	PERM.P.	OFF	TIMED	OFF
Y- R		5		10		5
Y- BG	42		42		70	

The timer chart identifies Y-R and Y-BG as the timer switches. The timer advances the dryer through its cycles, and determines whether the heater is ON or OFF.

Cross-hatching indicates the period for which the switch is closed; number indicates running time in minutes.

The Y-R switch is *open*, or off, for 10 minutes of the Permanent Press cycle—the heater is off during that period.

The malfunction might be anywhere on this circuit. Following the wire down from L1, you first come to a switch marked Y-BG. This is a timer switch; with the timer turned on, power can reach the other dryer components in the circuit. Test the timer with a multitester. (These procedures are described in the dryers chapter, page 114). Does the timer test OK? Then move along to the next component, the door switch, marked D-D1. When the dryer door is closed, so is the switch, allowing electricity through the circuit. Test the switch—does it work? If so, test the next component, a switch marked CO-NO, controlled by the start button. If there's no problem there, how about the next part, the drive motor? Use a multitester to test for proper resistance. If it passes the test, go on to the next component, a thermal fuse. Here the continuity tester light does not glow—you have found the problem. Replace the burned-out fuse, and the dryer will work—unless a faulty thermostat has caused the fuse to blow. (Find the thermostat symbols on the 240-volt circuit, locate the thermostats in the machine and begin testing them.)

Wiring diagrams for such 120-volt appliances as washers and refrigerators usually show only two incoming wires. These may be marked "L1" and "N," or distinguished by the color of the wires: black for L1 and white for neutral. A third wire might be marked "G" for ground. Sometimes wires in the diagram are labeled by their color, such as "Y" for yellow, or "PK-BK" for striped pink and black.

The timer chart *(page 138)* works with the wiring diagram to further pinpoint the source of a problem. In another example, you have set the dryer on Permanent Press, and find that your clothes are wrinkled and hot when the cycle is over. According to the sample chart on page 138, timer switch Y-R is supposed to be open for the last 10 minutes of the Permanent Press cycle, shutting off the heat and allowing the clothes to dry wrinkle-free. Could switch Y-R be malfunctioning? Test the timer; if the switch works, test the other components on the same circuit.

SAMPLE WIRING DIAGRAM FOR A CLOTHES DRYER

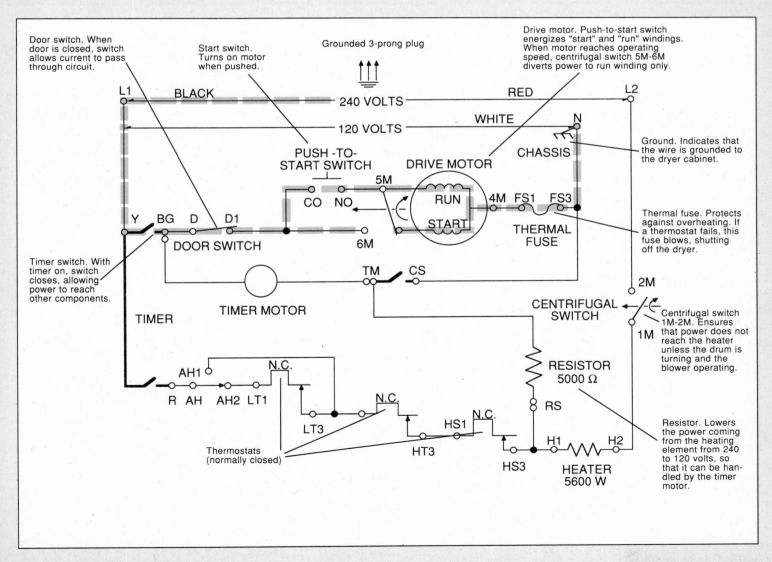

Door switch. When door is closed, switch allows current to pass through circuit.

Start switch. Turns on motor when pushed.

Grounded 3-prong plug

Drive motor. Push-to-start switch energizes "start" and "run" windings. When motor reaches operating speed, centrifugal switch 5M-6M diverts power to run winding only.

Ground. Indicates that the wire is grounded to the dryer cabinet.

Thermal fuse. Protects against overheating. If a thermostat fails, this fuse blows, shutting off the dryer.

Timer switch. With timer on, switch closes, allowing power to reach other components.

Centrifugal switch 1M-2M. Ensures that power does not reach the heater unless the drum is turning and the blower operating.

Resistor. Lowers the power coming from the heating element from 240 to 120 volts, so that it can be handled by the timer motor.

Thermostats (normally closed)

WORKING WITH GAS

Locating shutoff valves. Gas enters your home via a main supply line, passes through a meter that measures usage, then travels to gas appliances by means of galvanized pipes. A main shutoff valve near the meter opens and closes the gas supply to the entire house. Individual shutoff valves are usually located on the supply pipe at the back of each appliance. Shutoff valves vary in appearance, but all operate the same way. When the key or handle is parallel to the supply pipe, the valve is open; when it is perpendicular, the valve is closed. If you cannot move the valve by hand, use pliers or a wrench. If you smell gas and suspect a leak, open doors and windows, close the main shutoff valve, then call the gas company *(page 12)*. If there is a faint odor of gas, check to see that all pilots are lit, and check for a leak by brushing a soap-and-water solution onto pipe fittings *(page 126)*. If the solution bubbles, tighten the fitting or call for service.

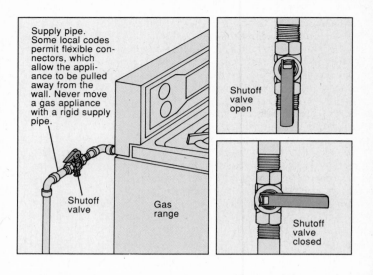

WORKING WITH PLUMBING

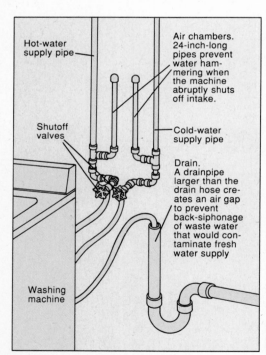

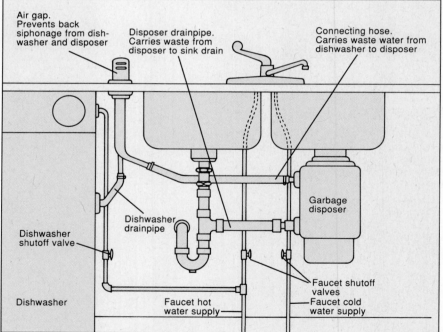

Water supply and drainage. From the large water main that enters your house, supply pipes branch out to carry water to appliances and fixtures. A shutoff valve, usually located between the water meter and foundation wall, allows you to stop all flow of water into the house in the event of an emergency *(page 13)*. Individual appliances have their own supply pipes and shutoff valves. Both the hot- and cold-water supply lines to a washing machine are fitted with simple gate valves *(above, left)*. The washer drain-pipe must be higher than the water level of the machine, or it will siphon water from the tub at the wrong time. A dishwasher is connected to the hot-water supply line of the kitchen sink. If the dishwasher has no shutoff valve of its own, its water supply can be turned off by closing the valve on the sink's hot-water pipe. Both dishwashers and garbage disposers share the sink drainpipe *(above, right)*. Many local plumbing codes require both appliances to be installed with an air gap that prevents back siphonage.

COSMETIC REPAIRS

Touching up a damaged cabinet. Even minor nicks and scratches can be unsightly when an appliance is in a visible location in your home. Enamel touch-up kits are available at hardware or auto-supply stores; for a special color you may need to ask the manufacturer or a parts-supply dealer. Before painting, remove any rust with fine sandpaper or the edge of a razor blade. Clean the area, allow it to dry thoroughly, then apply the paint with a small artist's brush. Deep nicks and scratches can be filled with several layers of paint. Wait for each layer to dry before applying a fresh coat.

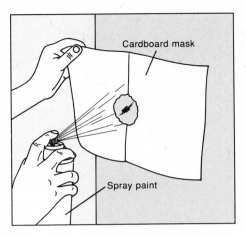

Cardboard mask

Spray paint

If a large area needs repainting, spraying is the best way to match the original surface. Sand the spot with extra-fine sandpaper, clean and allow to dry. Before spraying, ventilate the room and wear a spray mask. For optimal results, cut an irregularly shaped hole, slightly larger than the sanded area, from a piece of cardboard *(left)*. Hold the mask 6 inches from the appliance surface, then move it in a circle while you spray through the hole. Avoid excess paint, which could drip and run. Several light passes are better than one heavy coat.

GETTING HELP WHEN YOU NEED IT

To replace a broken part, first determine the appliance's model and serial number, stamped on a plate attached to the cabinet *(below)*.

If you can't locate the plate, consult the owner's manual. A diagram in the manual will also tell you the name of the part. Some mechanical and electrical components carry their own serial numbers, either stamped on the part or listed in the manual. Armed with this information, you can call the dealer to make sure the part is in stock. If possible, take the defective part with you for comparison.

Most dealers do not permit returns or exchanges, so make sure the replacement is identical—especially since an incorrect part could damage your appliance or, in the case of an electrical component, cause a fire. So-called "universal" parts are standard for most brands of an appliance. Before buying such parts, however, make sure they are approved for your machine. The parts dealer or distributor should be able to advise you. If you need more information, many manufacturers maintain toll-free "hot lines" to answer customers' questions. Look in your owner's manual for the number, or call 1-800-555-

1212 to find out if the manufacturer is listed. Complex mechanical parts such as transmissions or motors can sometimes be rebuilt rather than replaced, saving you money. Ask your dealer or distributor whether this is an option.

You have several sources of replacement parts from which to choose:

Independent appliance dealers. If a store does not carry the parts you need, the dealer may be able to order them for you or tell you where to find them.

Major retail chains. Hardware or department stores stock parts for the brands they carry, as well as universal parts.

Appliance parts distributors. These outlets usually sell parts to dealers and repairmen, but often welcome do-it-yourselfers as customers. If necessary, you may be able to order hard-to-find parts through the mail. Distributors usually carry service and repair manuals for appliances.

Manufacturers. Many do not sell parts directly to the public, but they may direct you to a retail source.

When it's time to call in a professional for repairs, choose carefully. Ask friends or neighbors for advice, or consult the retailer who sold you the appliance. The manufacturer or your local Better Business Bureau may also be able to recommend a reputable service technician. Make sure whoever you find is authorized to service your brand of appliance; if not, the repair may not be covered under the warranty.

When you phone to arrange a service call, be prepared to supply as much infor-

mation as possible. Know the model of your appliance, its age and condition. Take careful note of the problem: When did it occur (in which cycle or function), what sounds did the machine make, were there telltale odors, and is this a recurring problem? With these details, a repairman is more likely to bring the proper replacement parts. If your problem involves a flooded dishwasher or washing machine, you can save the repairman time—and yourself money—by bailing out the appliance before he arrives *(page 13)*.

Discuss the job with the technician before he begins. What rates does he charge? How long might it take? Will you be notified if the cost will exceed his estimate? Is the work guaranteed? Offer specific information concerning symptoms, and suggest your diagnosis, if you have one—without trying to tell the repairman his business. Once you've reached an agreement, leave him to his work. And before he goes, inspect the work carefully, and be sure you've received—and verified—an invoice itemizing parts and labor. Such a document will come in handy if the problem recurs.

You and your warranty. Before attempting any repair, check the appliance's warranty—and take note that some parts may be warranted for a longer period than that of the entire appliance. If the warranty is still in effect, you may void coverage if you do the work yourself and fail. Only an authorized service technician should make the needed repairs.

INDEX

Page references in *italics* indicate an illustration of the subject mentioned. Page references in **bold** indicate a Troubleshooting Guide for the subject mentioned.

ACKNOWLEDGMENTS

The editors wish to thank the following:
The Association of Home Appliance Manufacturers, Chicago, Ill.;
Jim Barcomb, Barcomb's Furniture, Plattsburgh, N.Y.; Jean Beaulé,
Gilles Bertrand and Marcel Rousseau, Gaz Métropolitain, Montreal, Que.;
Kenneth Bieber, Appliance Parts Distributors, Croydon, Pa.; Philippe Brault
and Gilbert Séguin, B & N Services, Montreal, Que.; Jim Buckler, Design
and Manufacturing Corporation, Connersville, Ind.; District of Columbia
Fire Department, Community Relations Unit, Washington, D.C.; Jim Evans,
Whirlpool Corporation, Benton Harbor, Mich.; General Electric Company,
Louisville, Ky.; Mike Hamilton and Jameel Khan, Montreal Appliance,
Montreal, Que.; Hechinger Company, Landover, Md.; Laplante and
Associates, Broan Ltd., Montreal, Que.; Clément Lessard and Richard
Martel, Canadian Appliance Manufacturing Company, St. Leonard, Que.;
Gordon McCord and Adrian Sato Jr., Mr. Service, Egg Harbor City, N.J.;
Richard Morgan, Waugh & Mackewn, Montreal, Que.; George Nardini,
Nardair Refrigeration, Montreal, Que.; Potomac Electric Power Company,
Washington, D.C.; G.N. Séguin, Montreal, Que.

The following persons also assisted in the preparation of this book:
Mary Ashley, Claire Dutin, Veronica Earle, Francine Lemieux, Barbara Peck
and Odette Sévigny.

Typeset on Texet Live Image Publishing System.